Resilient Cities

Re-thinking Urban Transformation

The RESILIENT CITIES book series aims to analyse the challenges faced by cities and provide an up-to-date body of knowledge, including a systematic collection of global cutting-edge best practices, fundamental to managing the urban transition toward resilience. The best practices will be collected and analysed following a common format, enabling the reader to understand the solutions adopted and clearly highlighting the parameters and possibilities for replication and up-scaling. The best practices are taken from a global city base including, Barcelona, Medellin, Adelaide, Copenhagen, Seoul, and Accra. The distinctiveness of the RESILIENT CITIES book series is its international dimension, coupled with a multidisciplinary and a cross sectorial approach. The RESILIENT CITIES book series will be a unique and fundamental resource for practitioners, policy makers and scientists involved in planning and governing the transition of cities. It presents the latest and up-to-date systematized information on research, practices and policies development, defining clear means and pathways for replication and up-scaling.

Octavio Francisco González Castillo
Valentina Antoniucci
Enrique Mendieta Márquez Margarita
Juárez Nájera
Alberto Cedeño Valdiviezo
Mariana Osorno Castro
Editors

Urban Resilience: Methodologies, Tools and Evaluation

Theory and Practice

recnet.
RESURBE

recnet.

SDU

Springer

UNIVERSITAT POLITÈCNICA DE CATALUNYA
BARCELONATECH
UNESCO Chair on Sustainability

unesco

Casa abierta al tiempo
UNIVERSIDAD AUTÓNOMA METROPOLITANA

Editors
Octavio Francisco González Castillo
Biological and Health Sciences Division
Biotechnology Department
Metropolitan Autonomous University - Iztapalapa
Mexico City, Mexico

Enrique Mendieta Márquez
Biological and Health Sciences Division
Department of Health Sciences
Metropolitan Autonomous University - Iztapalapa
Mexico City, Mexico

Alberto Cedeño Valdiviezo
Sciences and Arts for Design Division
Technology and Production Department
Metropolitan Autonomous University - Xochimilco
Mexico City, Mexico

Valentina Antoniucci
Department of Civil, Environment and Architectural Engineering
University of Padua
Padua, Veneto, Italy

Margarita Juárez Nájera
Basic Sciences and Engineering Division
Energy Department
Metropolitan Autonomous University - Azcapotzalco
Mexico City, Mexico

Mariana Osorno Castro
Metropolitan Autonomous University
Mexico City, Mexico

ISSN 2524-5988 ISSN 2524-5996 (electronic)
Resilient Cities
ISBN 978-3-031-07585-8 ISBN 978-3-031-07586-5 (eBook)
https://doi.org/10.1007/978-3-031-07586-5

This Springer imprint is published by the registered company Springer Nature Switzerland AG
The registered company address is: Gewerbestrasse 11, 6330 Cham, Switzerland

Foreword

Dear reader, the book you have in your hands, entitled *Urban Resilience: Methodologies, Tools and Evaluation*, is part of the Springer collection *RESILIENT CITIES: Re-thinking Urban Transformation*, which collects the initiatives of *Resurbe* program, which since 2014 has organized several international meetings (Spain, 2014; Colombia, 2015; Mexico, 2016 and Argentina 2016) with the intention of sharing and systematizing experiences and reflecting on the field of urban resilience, with a view to participating in the United Nations Conference on Housing and Sustainable Urban Development, *HABITAT III* that was held in Quito, Ecuador, from October 17 to 20, 2016.

Overpopulation, increasing urbanization, impoverishment and environmental deterioration represent current trends that we cannot fail to consider when we imagine the future of cities and with them, of humanity (since 2007, more than half of the world population lives in cities). At the same time, cities present various degrees of vulnerability to natural events (hurricanes, earthquakes, tsunamis, pandemics, among others) and human events (accidents and terrorism) that increasingly threaten their stability, as well as the life of their inhabitants and the integrity of its infrastructure.

We are experiencing a stage of history that Ulrich Beck (1986) identified as 'The risk society'. How can we prepare for risk scenarios in which natural and human phenomena impacts, each time frequently and with increasingly destructive potential, to socioecosystems? With the firm intention of joining the effort to approximate viable answers to these questions, this book brings together various contributions that include conceptual reflections and practical experiences (casuistic) around the topic of urban resilience, with a particular emphasis towards its methodological aspects, the use of tools and evaluation.

The complexity and multidimensional nature that characterizes the phenomenon of urban resilience determine that its approach requires incorporating the perspective of its various actors and relevant disciplinary fields. In the two volumes that make up this book, the experiences of 56 authors are presented; each of them develops their work as part of a team in 32 different institutions, located in 15 countries (Argentina, Australia, Brazil, Colombia, Denmark, Germany, Greece, Holland,

Italy, Ireland, Mexico, Portugal, Spain, United Kingdom and United States of America), from 3 continents: (America, Europe and Oceania); their professional fields cover a wide range of disciplines from: (a) the natural sciences (agronomy, biochemistry, biotechnology, chemistry, ecology and environment, geography, human ecology, physics), (b) the social sciences and humanities (anthropology, communication, economics, sociology, history), (c) design and engineering disciplines (architecture, artificial intelligence, chemical, civil, computing, design, electronics, environmental, industrial, landscaping, territorial planning, public policy, regional development, structural, systems, town planning, telecommunications, transportation), as well as various fields of mathematics, applied statistics and operations research.

We are in debt, and say thanks: to all authors, for their dedication, commitment and motivation to share their experiences; to the organizers of the Series: Nicola Tollin, Jordy Morató and Ernesto Santibanez, for their great vision and capacity of convocation to promote this initiative; to the academic team: Valentina Antoniucci, Enrique Mendieta M., Margarita Juárez N., Alberto Cedeño V., without whose committed work this book would not have been possible; to Mariana Osorno C., for her valuable participation in the revision and adjustments during the final integration of the materials; and to Margaret Deignan, Marielle Klijn, Nagarajan Paramasivam and all Springer editorial team.

Being the city, the focus and frame of reference from which the different experiences of this book are developed, it is not surprising that in each of the chapters some lines are included to locate, size, and reflect on the urban phenomenon and its growing expansion on the planet. On the other hand, since resilience is the attribute of the urban that attracts and concentrates the attention of the authors, it is not surprising that several pages are devoted to presenting, defining, and contextualizing this important property of systems. It should be mentioned that the field of resilience – like that of urban sustainability to which it contributes – represents a theoretical-methodological territory under construction and continuous reconstruction. Both fields attract the attention of theorists and practitioners from very different perspectives around the world. This places us in front of a puzzle in process that continually receives new pieces (conceptual, theoretical, methodological, or casuistic) that must be valued and accommodated until the field is consolidated.

Mexico City, Mexico Octavio Francisco González Castillo

Padua, Veneto, Italy Valentina Antoniucci

Mexico City, Mexico Enrique Mendieta Márquez

Mexico City, Mexico Margarita Juárez Nájera

Mexico City, Mexico — Alberto Cedeño Valdiviezo

Mexico City, Mexico — Mariana Osorno Castro

Odense, West Yorkshire, Denmark — Nicola Tollin

Terrassa, Barcelona, Spain — Jordi Morató

Preface[1]

According to the World Bank, more than half of the global population lives in cities, and it is expected to increase to two thirds by 2050. This dynamic by itself increases the challenges and complexity of urban regions, and the overall instability of climate adds new threats, many of which remain poorly understood.

Urban resilience is currently an open research field, for which there is no consensus or mainstreamed approach yet; it is thereby necessary to build new conceptual approaches, rethinking the principles of urban resilience including ontological, epistemological, methodological, and axiological assumptions. The crisis of sustainability and resilience for cities presents the following complications: (a) multiple and diverse systems with a complex interaction, (b) multiple dimensions and levels of analysis, (c) multiple stakeholders and institutions which have different worldviews. Transition to urban sustainability requires new strategies, decision support systems, methodologies, tools, and evaluation procedures for multi-level governance, as well as integration of risk management, mitigation, and adaptation, within a self-adaptive, proactive, and participatory frame. Thereby, a transdisciplinary systemic approach for urban resilience, in the wider frame of sustainability paradigm, is needed to implement synergic transformative actions.

There is a need to increase the knowledge base and strengthen the science-policy interaction, to properly assess the urban-regional metabolism and increase the socioecological sustainability and resilience for face global challenges, there is a need to enrich the knowledge base and strengthen the science-policy interaction. The co-creation of an accessible and transparent base of knowledge can lead to a better understanding of the urban socio-ecosystems' strength and weaknesses, as well as to raise awareness of its diversity, creativity, and social capital.

There are currently no consistent sets of indicators able to fully tackle the multidimensionality of urban resilience; thereby, it is necessary to further develop

[1] This preface takes, as starting point, the call with which those of us who make up the initial team – Luis Jiménez Herrero, Josep Pont Vidal, Astrid Rocuts, Elena Perez Laguela, Jocelyn Parada Ayala and Octavio González Castillo – gave shape to the book initiative and summoned to those who now are the authors of this book. We express our appreciation to all of them and say with satisfaction that the initiative that we promoted together has come to fruition.

resilience indicators and prospective models, which can reinforce risk management and generate opportunities. Urban resilience evaluation depends on a coherent understanding of the underlying concept of risk and the socioecological interactions of urban and regional systems, which rely on natural capital to implement nature-based solutions.

Measurements and evaluation tools, including new information technologies, geographic and conceptual information systems, are essential to assess the impacts, risks and vulnerabilities derived of climate change. Resilience and disaster risk reduction assessment should be part of any comprehensive urban sustainability strategies. This volume compiles conceptual works and case studies related which methodologies, tools and evaluation frameworks that are deemed useful to guide urban transition towards resilience and sustainability. The topics with which the authors were summoned were:

T1: Urban resilience approaches and paradigms. Given the uncertainty and risks associated with global change and climate change that cities and regions face, the concept of urban resilience arises to address systemic challenges and guide the urbanization process within the paradigm of sustainability.

T2: Complex systems methodologies for urban resilience and sustainability, including multi, inter and transdisciplinary approaches. A systemic vision and inter/trans-disciplinary approaches for urban resilience are necessary to complement 'normal science' and to implement synergic transformative actions. Conceptual advances in inter/transdisciplinary approaches are central to analysing urban sustainability and resilience, and to develop solutions to the complex problems of urban systems.

T3: Tools, indicators, and modelling for urban resilience. This topic aims at advancing the measurement and evaluation systems of urban resilience and sustainability through the development of indicators and dynamic models that better reflect the complex interactions and the relations of cities and regions. The base of knowledge should be increased to boost the 'science-policy interface' in decision-making. This also includes new approaches to development indicators, as well as forward-looking models to assess urban resilience policies and support governance systems at the bioregional level.

T4: Geographic information systems and conceptual information systems. Cartography allows the development of maps and information systems guiding both geographic and conceptual exploration that can facilitate the communication among experts, in the frame of complex and multidimensional contexts, while enabling the identification of conceptual regions that have not been explored or developed yet.

T5: New approaches in evaluation in context of urban resilience. Conventional approaches for evaluation do not appear to be contributing sufficiently to the achievement of urban resilience, thereby it is necessary to build a conceptual platform based on a new set of ontological, gnoseological and axiological assumptions that can facilitate the development of the new approaches of urban resilience.

Mexico City, Mexico Octavio Francisco González Castillo

Introduction

It is through a constant interaction with the environment that the human being is continually getting to know, reflecting, intervening, and adapting the environment, and himself, to the reality that surrounds him. That's the way human beings create and recreate their conceptual world – that of ideas and thoughts – and draw bridges between this and the 'real world' that surrounds them. Regarding the formalization and construction of these conceptual-theoretical-methodological frameworks, belonging to the conceptual world, González-Castillo (2008, p. 129) states:

> *Personal knowledge represents the minimum functional unit of the individual thought born from cognitive research; it is a set of statements (constructs, definitions, and propositions) with a representative function from some part of the real world. When knowledge is publicly verified it becomes recognized as theory ... Personal procedure is the minimum functional unit of individual thought born from the design of intervention. It defines the sequence of activities to be carried out to achieve an end. When a procedure is publicly validated it becomes recognized as a methodology ... personal reflection is the minimum functional unit of individual thought born from reflective introspection, which is considered as a set of statements (constructs, definitions, and propositions) with function representative of a part of the conceptual world. When she is publicly recognized, she becomes considered a reflexive.*

According to Javier Navarro, the idea of casuistry is applicable in all those contexts in which an attempt is made to understand a global situation based on some specific cases that are illustrative. For Miller, casuistry is a genre of research that addresses specific problems, their correct interpretation, and their reasonable resolution. For her part, Guerra-Palmero M.J. (2013, p. 10 and 13) affirm:

> *New casuistry does not offer recipes, but defends the need for analysing, pondering and estimating circumstances... because every rule to get along with the complexities of the human requires exceptions ... if not, at least, of a certain work of interpretation that takes into account the concrete details and the demands of the context, a certain hermeneutic that makes the application of the principles more flexible to adapt them to the diverse cases. The procedure will be to locate in the case in question a set of cases with which it maintains similarities. Analogous cases, by proximity, will illuminate the case study in a kind of search for precedents. By using the analogy, both differences and similarities will be determined. These linking cases are the essence of casuistic proceeding, as opposed to the isolated consideration that seeks references directly in the principles. In this way of*

> *proceeding, a series of type-cases or paradigmatic cases are accumulated that serve to learn, deliberate and gradually systematize the analytical experience. This systematization of paradigmatic cases allows us to point to a series of relevant maxims and rules.*

In each chapter of this book, the reader will find a particular amalgam of conceptual, theoretical, methodological, and casuistic contributions around urban resilience with an emphasis on its instruments, methodological and evaluation aspects.

Although each chapter harmonizes with the whole that makes up the book, they also maintain their individuality and are self-contained. For this reason, readers may decide to start their reading through any of the chapters and continue with that freedom for the rest of the book. However, if you decide to follow the order in which they are presented, you will find a natural progression:

- *Chapters 1 and 2* introduce us to the notion of urban resilience, as well as the various approaches that have emerged to approximate the phenomenon.
- In *Chapters 3, 4 and 5* case studies are offered, each from a different perspective (landscape architecture, urban planning, and urban sociology).
- *Chapters 6 and 7* deepen the knowledge of urban resilience, establishing both its differentiation with respect to other areas of regional resilience, and its link with concepts such as 'balance', 'stability' and 'sustainability', 'fragility' and 'antifragility'.
- Finally, *Chapters 8, 9, and 10* take up the case study to study the phenomenon of resilience at three different levels: urban socio-ecosystem, county, and eco-neighbourhood.

The following is a brief description of the context for each of the ten chapters that make up this first volume of the book.

Chapter 1: *Searching a Resilient City: A Study About Theoretical-Conceptual Joints Between Smart City and Urban Resilience*. From the perspective of urban planning, the authors place us in front of the current trend to use more and more technology, particularly ICT (hardware and software), to attend to the daily and emerging needs in cities. The authors question: what relationship exists between the 'smart city' and 'resilient city' models? Does advancing in the direction of the first necessarily imply moving in the direction of the second? How different conceptual approaches to smart cities contribute to enrich the debate on the improvement of urban resilience? Looking for answers to these questions, the authors identify, through an extensive bibliographic review, the root or founding concepts in both fields for then submit them to a conceptual correlation study. This is how the authors invite us to reflect on (a) whether one concept contains or complements the other? (b) the points of closeness and distance between them and (c) the need to conceive new concepts in urban planning: theories and models. The authors issue a warning about the great diversity of approaches (academic, institutional and market) that converge in the definition and deployment of both concepts. By locating it as a field in which economic and political interests compete, the authors invite us not to lose sight of the fact that the ultimate goal must be people's 'quality of life' and not to be carried away by superficial and fashionable definitions derived from a reductionist tendency that overestimates the role of technology (technocentrism) for the design and operation of cities. As the reading progresses, the chapter explores and reveals the semantic field of many of the terms through which, later, the various authors of the book will gradually develop the theoretical-methodological field of urban resilience.

Chapter 2: *Framing 'Resilient City': Systemic Versus Community Focussed Interpretations of Urban Climatic Resilience*. Using the notion of 'paradigm', the author seeks to make explicit what characterizes, and at the same time makes different, two approaches that aim to guide urban resilience: (a) systemic: engineering perspective oriented by urban infrastructure and (b) communitarian: sociological perspective, oriented by culture and of social action patterns. The chapter begins by contextualizing the risks for the urban environment associated with climate change and then introduces the notion of resilience as useful to face it. However, it warns us that being a multi-factorial and complex property, the implementation of actions that seek to increase resilience has the potential to impact, positively or with collateral damage, the daily life of citizens, as well as various urban dimensions (aesthetic, spatial, of infrastructure, economic and business, among others). To characterize both paradigms the authors initiate a reflection aimed at answering the following questions: What is the problem? For whom is this a problem? What are the most important causes of this problem? What important values are being threatened? What should be done and by whom? Throughout the chapter, it will be clear that different paradigms will lead to the selection of different aspects of a perceived reality and promote particular definition of the problems, causal interpretations, moral evaluations, and action recommendations for solving problems. It is due to this power implicit in the paradigms that when they are unconsciously used by the various urban actors and agents, the negotiation can lead to a 'dialogue of the deaf' in which the most powerful or persuasive actors impose their points of view and particular interests. The author concludes by inviting us to use both 'frames' in an eclectic and dynamic way, maximizing synergy and reducing disadvantages among them in each city intervention particular situation.

Chapter 3: *Townscape Catalogues Toward Urban Resilience and Sustainability*. The authors place their reflection in a scenario in which urban planning and its agents are affected by deficiencies and limitations that have made it difficult to stop and reverse current trends in which the carrying capacity in socio-ecosystems is continually exceeded by impacts of the activity of the human being. This lack or inadequacy of urban planning threatens the landscape image and, therefore, endangers cultural-natural heritage and environmental balance. Against this background, the Townscape Catalogue is presented as an urban planning-management instrument developed with an orientation towards caring for the environment and well-being of the citizenry. Assuming a transdisciplinary view of the complex systems, the analysis of the landscape is taken as the core framework for integrating various sources of information: field observation, aerial photography, satellite analysis, historical documentation, socio-environmental statistics, surveys, and interviews. The Townscape Catalogue is developed in three stages: diagnosis, characterization, and evaluation; in these, the urban landscape is studied through its tangible and intangible components, assessing its value, dynamics, and quality condition, which allows to classify its various regions as vulnerable, protected or with opportunity and consequently propose actions for its preservation, recuperation, and enhancement. In this way, the Catalogue may entail a path to the construction of 'flexibility and resilience patterns' of a part of the city; 'buffer' sectors for the recovery, mediation, and endurance of systemic urban impacts; and the detection of landscape components which are apt to articulating the organic operation of flexible urban areas. This is how the Townscape Catalogue becomes a tool for the analysis and feedback between instances of designs and transformation of the city. The initiative is illustrated through the case study of Suquía River in Córdoba, Argentina.

Chapter 4: *Building Resilient and Sustainable Cities Starting from the Urban Transport System.* As the chapter progresses, the underlying complexity of urban resilience phenomena is showing by the authors: the complex nature of operations that support daily life in urban systems not only generates the potential for unforeseeable failures and cascading effects, but it also creates unexpected opportunities for acts of disturbance, as well as terrorism. Given this scenario, an interdisciplinary experience is presented that, under the 'resilience engineering' approach and the use of ICT, proposed the design and implementation of an information management system that supports decision making, in real time, to strengthen resilience of the Urban Transport System (UTS) in Florence, Italy. In the words of the authors, currently UTS have to cope with significant vulnerabilities: ageing infrastructure; multiple relevant dimensions, actors, metrics, and decision makers; increasing concentration of populations; interdependencies among the physical and cyber infrastructures; multimodal transport stations with potentially hazardous facilities; growing threats of climate change and terrorism. The challenges to UTS resilience reside then in the ability to monitor and control the available resources and the capacities to coping with performance variability, in expected and unexpected scenarios. The Application Framework is aimed at defining indicators (operationalize resilience parameters), create guidelines, methods, and tools to manage the phases of urban resilience, namely: preparation, reaction, adaptation, and learning. To achieve this, it is necessary to implement models, algorithms, and rules for data mining and analysis; predict network evolution; learn and anticipate events that could lead to a failure and communicate those to the monitoring and actuation components, thus ensuring an adequate response (following predefined guidelines) to disturbances. The case of UTS on Florence, Italy, generously illustrates the complicated reliance that urban resilience has on the infrastructure that supports and provides services to the city.

Chapter 5: *Creating a Resilient City: A Community Focused Approach in Bogota, Colombia.* The author changes the 'systemic' approach, dominant in previous chapters, and places us in front of a "community" one, from which he approaches urban resilience with an ethnographic, holistic, and transdisciplinary perspective. Taking as a starting point a general notion of resilience: 'Ability of a biological being, system, structure, person, or community, to positively adapt and bounce back despite significant threat or stress' the author presents the case of two vulnerable communities in the city of Bogotá, Colombia: (a) the community of a Lesbians, Gays, Bisexuals and Transgender (LGBT) Care Centre, and (b) a community served by the District Institute of the Arts (IDARTE). In both cases, the results of contrasting the discourse analysis of the public policy document "Plan de Desarrollo, 2012–2016: Bogotá Humana" against a series of semi-structured interviews applied to diverse stakeholders are shown. The chapter emphasizes the fact that although vulnerable groups are subject to environmental risks such as those associated with climate change, they are also exposed to the onslaught of disease, marginalization, rejection, discrimination, and isolation, as well as intolerance, violence, and indifference. Looking answers for the question: how government can build the necessary tools to contribute community resilience? the author applies a series of ethnographic techniques such as participatory observation, interview, and discourse analysis with the intention of identifying and promoting public policies that increase the level of security, wellness, and resilience of vulnerable groups in urban areas. The author concludes that participation culture, community integration, funding, and social programs are factors that strengthen the community resilience of these groups, while poverty, iniquity, disarticulation, and social indifference are factors that weaken it.

Chapter 6: *Resilience and Sustainability in Urban Socioecosystems: A Conceptual Reflection*. This chapter seeks to put the urban phenomenon, as well as the search for its resilience and sustainability, in the context of the current civilizational crisis. The deep causes of the crisis have their roots in the history of the Anthropocene and extend its branches to the 4.5 million years of the humanization process followed by the homo genus. Although the biological and cultural progress achieved by humans is surprising, in what we can consider as only an 'instant' in the history of planet Earth, the progressive appearance of phenomena such as pollution, loss of species, global warming, hunger, disease, poverty and violence has emerged as anomalies of a civilizing model and triggered, since the middle of the twentieth century, an awareness of deterioration as the 'other side' of the development coin. The magnitude of the human impact on the planet makes it no longer possible to follow the path of development without taking control of its counterpart of deterioration. To face the crisis and transform it into an opportunity, the human being must accept the challenge of reviewing those models of thoughts and paradigms that from his 'conceptual world' (noosphere, social imaginary, worldview, culture) have guided him in his cognitive approach, reflection, and intervention on the 'real world'. The rest of the chapter flows to answer the questions: What purposes has urban development served, under an anthropocentric approach? What new purposes should aspire to serve if development assumed a socioecocentric approach, oriented towards the resilience and sustainability of socio-ecosystems (wild, rural, and urban)? The author invites us to make the local and global crisis the starting point for the construction of a new development approach, one in which the final beneficiary for the development should be the socioecosystems. To achieve this, the author concludes, it will be necessary to first change our cognitive and intervention approaches, as well as design novel methodologies and instruments to assess resilience and sustainability in socioecosystems.

Chapter 7: *System Approach to Resilience-Based Design: Political Decisions and Steps Towards Antifragility*. The chapter begins with a warning from its authors: variability and randomness are rules and not the exception in everyday life... it is better to live and deal with it, and try to take advantage by using, in a contingent way, various strategies. A review of catastrophic events that took place during the first two decades of the twenty-first century has shown us the fragility that urban systems have in the face of extreme phenomena of both natural and anthropogenic origin. Frequency and magnitude of these catastrophic events have contributed to making evident the need for improving models and tools for evaluating and managing "performance" and "resilience" of urban systems. Nevertheless, one issue remains to be solved: there are no standards, and the needed metrics are difficult to implement. Facing this challenge, authors' motivation is to help advisers and experts in decision-making of public policies to address the consideration of the different risks that influence resilience in urban areas and therefore in well-being of society. With this statement, the authors begin a search for new paradigms of urban design and planning. They found that the innovative concept of antifragility (Taleb, 2012) – the ability of a system to benefit from the variability of the environment when the latter goes beyond a certain pre-established threshold – has the potential to contribute to facing urban threats or risks. Based on this powerful concept, the authors provide insight, framework, and a set of good practices advises on: (a) system modelling —hazard categories (independent, interacting, and chains of hazards) and time horizons for the analysis (historical, previous, during, aftermath, and log run)—; (b) political decision representation; and (c) data collection. Along with antifragility, the chapter explores other innovative trends as well as emerging instruments in the field of urban design.

Chapter 8: *An Integrated Methodological Framework to Assess Urban Resilience*. Assuming precautionary principle, the authors question the ability of the 'normal' (disciplinary oriented view of the positivist science) model to, by itself, approximate and face the complexity that characterizes today to the risks and challenges facing humanity. Instead, they promote 'post-normal' (sustainability sciences) model as a complementary alternative. In such direction, they propose a systemic framework to assess resilience in urban and other types of socio-ecological systems that combine the use of indicators, thresholds, dynamic simulation, as well as the assessment of policies, scenarios, and effects of uncertainty. The pretention of the approach proposed is overcome limitations of conventional Decision Support Systems (DSS) and traditional risk analysis that frequently are based on the assigning of weights to each criterion, which are then combined into mixed indexes to find an optimal solution. The approach proposed avoids the use of indexes mixing non-reducible dimensions and, instead, keeps track of the positive and negative effects of each policy or scenario on environmental, economic, social, and institutional factors within the urban concerned as well as their interactions with other local or global socio-ecological systems involved. The establishment of threshold values for each indicator provides a way to identify and assigned the lowest priority – even reject – those policies that would exceed the concerned resilience thresholds. The participation of involved actors, particularly in the first stage (definition of objectives and indicators) and in the final stage (assessment of policies and scenarios), is essential to build confidence in the indicators system and to increase the stewardship of policy makers and stakeholders in the achievement of the objectives. To support and illustrate proposal, three practical cases are presented: (a) Galapagos Islands, (b) Murcia City and its periurban agro-ecosystem, and (c) Fuerteventura (Biosphere Reserve).

Chapter 9: *Sustainable and Resilience Descriptors for the Xochimilco-Tláhuac Lacustrine Area at México City*. The chapter present a case study that highlights the methodological relevance of build a 'powerful gradient' between (a) description of the current problematic scenario (diagnosis) and (b) goal image or idealized scenario designed for the study-intervention system. The 'gradient' creates a creative potential to identify strategies capable of moving system either: (a) away from the problematic situation (resilience and sustainability low profile) or (b) closer to the idealized situation (resilience and sustainability high profile). The study took place in a polygon located between Xochimilco and Tláhuac, two counts of Mexico City: floodable conservation area that since 1987 takes part of the World Heritage by UNESCO. The polygon is also part of the 'Green Horseshoe', a rural-urban transition region, shaped like half-belt green that borders the southern part of the Metropolitan Zone of the Valley of Mexico, one of the ten largest, complex, and populous urban areas in the world. The project was part of the program Sustainable Integral Metropolitan Zone that in the first decade of the twenty-first century was led by the Ministry of Urban Development and Housing of the Government of Mexico City and integrated an interdisciplinary group of researchers from the Autonomous Metropolitan University with the objective of design and evaluate future scenarios for the Polygon. Based on the Plus Planning Cycle methodology, which integrates experiences from 38 cities and regions affiliated with The Plus Network, the authors developed descriptors, grouped into five types of environments: natural, built, economic, social and government, which were initially used as a framework to describe the current situation and the prospective image of the polygon, to subsequently develop action guides aimed at leading the polygon towards a scenario of resilience and sustainability.

Chapter 10: *Eco-neighbourhoods as a Form of Urban Resilience*. The chapter describes how eco-neighbourhoods began to emerge with the global push for sustainable development in the 1990s and highlights their contribution to urban resilience. The author recovers the criticisms of Jane Jacobs, who in his book Death and Life of Great American Cities (1961) wrote about the mistakes of modern urban planning, that have turned neighbourhoods of cheap housing into a place of crime and social despair, and claims that they undoubtedly, in the motivations of the crime there is a substrate of deep and complicated social pressures. Eco-neighbourhoods, state the author, represent a possibility of a resilient and sustainable city, and could provide an example for local governments as well as for inhabitants in the metropolitan areas. As the chapter advances, the author develops concepts as eco-efficiency, habitable city, bioclimatic urbanism, eco-urbanism, eco-neighbourhoods, resilience, and sustainable city. At the same time, an extensive historical review is presented, as well as the description of internationally renowned eco-neighbourhoods, among others: Puchenau Garden City in Linz, Austria; Solar Village No. 3 in Pefki-Lykovryssi, Athens, Greece; Le ZAC du Raquet in Doaui, France; Viikki in Helsisnki, Finland; Vauban in Freiburg, Germany; Bedzed in England. As the casuistry is presented, a broad set of principles, criteria, instruments, and factors that act 'for' or 'against' resilience and urban sustainability are revealed. In concordance with Ester Higueras, and many other pioneers on the subject, the author mentions the following actions implemented in ecological neighbourhoods: closing the cycles of matter and energy; promote identity and social participation; the rehabilitation of the built and cultural heritage; compact and flexible design; connect public spaces to infrastructure; guarantee safe, friendly, and healthy public spaces; and give preference to sustainable mobility.

Finally, we want to share with you, dear reader, that the motivation that led the different authors of this book to share their experiences with you is not satisfied with the presentation of these, but rather aspires to motivate you to start your own search and commitment towards a resilient and sustainable urban development.

February 5, 2021 Octavio Francisco González Castillo

References

González-Castillo, O. (2008). *Design of a planning instrument to incorporate, in the evaluation of systems, approaches towards sustainability*. Thesis to obtain the doctoral degree. Faculty of Engineering, National Autonomous University of Mexico, Federal District, Mexico.

Guerra-Palmero, M. J. (2013). Casuistry and moral reasoning. In R. T. Hall, & J. S. Arellano (Coordinators). *Casuistry: A methodology for applied ethics*. Ed. Fontamara. Autonomous University of Queretaro.

Taleb, N. N. (2012). *Antifragile: Things that gain from disorder*. Penguin.

Contents

Chapter 1
Searching a Resilient City: A Study About Theoretical-Conceptual Joints Between Smart City and Urban Resilience

Tharsila Maynardes Dallabona Fariniuk, Alexandre Hojda, and Marcela de Moraes Batista Simão

Abstract The relationship between cities and technologies has been developing over the centuries. The complexity of urban space, associated with population growth and the use of natural resources highlighted environmental, socioeconomic and spatial vulnerabilities. In this scenario, some technologies are now being developed to make the urban space more resilient and smarter. The recent spread of the term "smart city", to designate cities that using technologies to assist in solving demands, suggests thematic approaches to urban resilience. This relationship, however, is not obvious and requires discussions on how these concepts are articulated and which aspects of smart cities make such cities more resilient. Considering this challenge, this paper aims to examine the relationship between the concepts of smart city and urban resilience, and asks how different conceptual approaches to smart cities contribute to enrich the debate on the improvement of urban resilience, focusing on studies about urban planning. To answer this question, we developed a correlational research, qualitative and documentary, which started with the search for convergences between the concepts and the construction of a frame from each one's fundamental characteristics. It was observed that although the link exists academically, the concepts can be worked under different approaches. As such, the joint consideration is important and can contribute to more solidified, efficient, and participatory urban environments.

T. M. D. Fariniuk (✉)
Unifacear, Curitiba, Brazil
e-mail: tharsila.fariniuk@unifacear.edu.br

A. Hojda
Nitzor Company, Jerusalem, Israel
e-mail: profalexhojda@gmail.com

M. de Moraes Batista Simão
City of Tampere, Tampere, Finland
e-mail: marcelamoraes.b@gmail.com

O. F. González Castillo et al. (eds.), *Urban Resilience: Methodologies, Tools and Evaluation*, Resilient Cities, https://doi.org/10.1007/978-3-031-07586-5_1

Keywords Smart city · Urban resilience · City and technology

1.1 Introduction

The advent of new infocommunicative technologies – the Information and Communication Technologies (ICTs) – particularly in the 1990s, initiated many social and economic transformations that are reflected in urban areas. These areas began to incorporate elements previously unknown or little used, such as sensors, cameras, georeferencing systems, wireless networks (WiFi), etc. From this moment on, several towns start to regard technology as key instruments of urban integration and as auxiliary instruments for actions and public services.

Recently, the term "smart city" emerged, to designate urban environments that use the technological potential to answer to demands, solve problems and enhance urban characteristics, also with trying to offer better public services focused on efficiency and welfare to citizens.

The collective imagination tends to relate the cities "smart" to the futuristic images that are conveyed in books and science fiction movies. The literature, however, seeks to conceptualize smart cities from the consideration of several key factors for the quality of urban life.

In the 2000s, particularly during the second half, the "smart city" expression became widely used by private technology companies such as IBM, Cisco, Siemens, etc. These companies associate the word "smart" as a kind of "product", which serves as an urban solution. These packages "pre-built" started to be offered to local governments with the promise that technology is the only solution ready to answer to urban demands. Thus, the discourse of "automatic city" became sold as the future of cities (Lazaroiu & Roscia, 2012; Hall, 2000). In this process, the labeling "smart city" was expanded in projects for different areas of social interest. As an example we can mention the Operations Center Rio de Janeiro (COR), Smart City Wien, among others.

Considering these factors, the smart cities should not be envisioned as futuristic cities, but as complex urban networks that can be monitored, regulated and managed with the use of technologies (Kitchin, 2014). Therefore, they are cities that can be more efficient in resource management (human, technical, material and institutional), the decision-making process and transparency, by exploiting the potential of technologies (Weiss et al., 2015).

This conceptual amalgam has become common in urban planning area, associating the concept of smart city to others thematics, such as sustainability and governance, considering that many factors can contribute to make a "smarter" city. One of the concepts which is most often associated with smart city is urban resilience.

According to Souza and Flanery (2013), a resilient city results from the action of different groups of actors (citizens, management, planners) in processes that mitigate urban vulnerabilities. This city model must be strong and flexible (Godschalk, 2003). Holling et al. (1995) points out that one of the main virtues of a resilient city is its capacity and ability to absorb natural disasters. Considering this, the author shows that a resilient city is not one in which critical situations never occur, but one that has the ability to return to normal, post-event state, as soon as possible.

Although the concept of urban resilience is often associated with the adaptation and restoration of natural and environmental phenomena, there is a conceptual convergence with the objectives of creating "smart" environments: mitigation and/or resolution of urban problems in a fast, flexible and transparent manner.

This process occurs because the intelligent city uses ICTs infrastructure to boost management efficiency. The process of data collection and analysis generate more knowledge, identifying patterns and relationships, and helps avoiding or minimizing problems related to critical situations (natural disasters, mobility, public order, security). Thus, the city becomes more responsive (Steenbruggen et al., 2015).

Considering these points, the question is: how different conceptual approaches to smart city contribute to enrich the debate about the improvement of urban resilience, focusing on studies about urban planning?

In this sense, this research aims to identify conceptual links between the models of resilient city and smart city, looking for:

(I) Comprises the logic of both processes.
(II) Investigate whether one of these concepts is contained in another or if they are complementary.
(III) Identify points of closeness and distance.
(IV) Envisage the application of these concepts in the study of urban planning area of theories and models.

1.2 Methodology

This study is part of a broader research project which aimed study the smart cities considering the relationship between people, technologies, spaces and institutions. In this main research, the concept of resilience is treated as a result of a process that, among other things, improves management through the use of strategies to "make the city smart." It demands, therefore, a more detailed observation of the elements of the concept of resilience and the factors that influence this process.

This research discusses how the concept of smart city enriches the debate on urban resilience. We proposed a conceptual theoretical survey of each concept to identify, later, joints between the concepts. This survey is based on a bibliography research and is characterized as a exploratory and correlational study.

The correlational study seeks to interrelate two different variables (in this case, smart cities and urban resilient) that one intervenes on another. In this type of study there is no assumption of relationship between the two variables before watching them – what happens in experimental studies, for example (Polit & Beck, 2011). The correlational research allows us to study the variables simultaneously, making it possible to identify what is the degree of connection between them. This type of research, therefore, allows the identification of more than just the presence or absence of an effect (as in experimental processes) (Gressler, 2004).

To make the correlational study, we first chose to seek convergences within each concept. Thus, after reading and analyzing literature, we elaborated a frame with four key features to the concept of smart city and ten key features to the concept of resilient city. These frames are presented at the end of their conceptual chapters.

After this definition, the total of 14 elements were arranged in a correlational framework, from which we discuss the joints. Then, it was made the analysis of correlations between the two concepts in the debate on urban planning.

1.3 Smart City

The concept of smart city is based on the use of technology in the city and society. However, although it used in the literature with different nomenclatures, context and meanings, the term still presents a high degree of abstraction and suffers from a lack of precise definition. There are cities and projects that call themselves smart cities with diverse approaches (environment, economy, infrastructure, online public services, etc.), and most of these projects aim to improve the efficiency of the systems that constitute the city (water, waste, energy, mobility, emergencies, e-government), with availability of information and offer faster and more assertive responses.

According to Weiss et al. (2015), the concept of smart city emerges as an evolution of the Digital City concept, from the improvement of various elements such as online data systems, intelligent architecture and power management. According to the authors, smart city is the city with more efficient in human, material and technical resources management, by means of the opportunities generated by the evolution of ICTs. Thus, there is a greater support for decision-making, democratization and transparency.

Nam and Pardo (2011) present the smart city as a joint environment between technology and the human enhancement. According to these authors, the combination of these two factors should result in quality of life for the citizens, with better offering in urban services, (re)design of processes and improvement of the ability to govern. In this line of thought, Klauser et al. (2014) believes that the smart cities based on the relationship between technology, human capital and institutional relations.

Mitchell (2007) correlates the smart city with the functions of the human body, emphasizing the combination of the technology with networks, sensors and integrated urban infrastructure, symbolizing the human body structures such as veins, channels, arteries, and the importance of proper functioning, the integration of organization and care (and maintenance) that should be taken with the city management. This author also states that the new "intelligence" of the city stay in the combination between physical, technological and information elements.

According to Calzada and Cobo (2015), there is a need to overcome the purely technocratic and individualistic concept of smart city, looking for greater consistency of articulation between what is individual and what is collective. According to the authors, the smart citizen, with the spread and intelligent use of information, may have more access to collaboration and transparency. The citizen becomes to be more interested in participation in decision-making.

In this sense, Nam and Pardo (2011) point out that the success of a smart city project is not determined by technology, but also depends on the quality of leadership and inter-organizational coordination. In the same line of thought, Citrigno et al. (2014) argue that the key to urban intelligence is the human factor, in other words, the consideration of citizens and involved workers are like "guardians" of the territory. These actors should have the ability to detect critical urban situations, acting as "social sensors." According to the authors, therefore, ICTs are not the fundamental part of the process, but it suits as auxiliary tools for the parts connections and promotion of collective urban intelligence.

Kitchin et al. (2016) establishes some reservations about smart cities concept. They highlight concerns about the use of the term: (a) it is a tendency to reductionism and to the thought that technology can solve any urban demand; (b) the tendency to corporatism (public institutions selling services) that can generate situations of dependence and monopoly; (c) the tendency to characterize impartially the use of data, without recognizing the fact of impact and barriers of ideologies and institutions.

Other relevant point is the comprehension of smart cities should go beyond from the scenarios about software and hardware, and should consider the fine line between development and environmental priorities (Luque-Ayala & Marvin, 2015).

Sustainability and related concepts have been receiving greater attention with regards to the smart city concept, and has led to the improvement of the term. Some authors call the new smart city as "sustainable smart city." According to Al-Nasrawi et al. (2016), this model articulates the notion of urban development to e-government practices and environmental policies in order to make the environment more and more "sensitive" to the context.

In this sense, Vanolo (2014) argues that the construction of a sustainable city is essential to consider the intelligence of urban space. Ratifying this argument, the authors Neirotti et al. (2014) argue that the increase of urban complexity, with the introduction of new technologies, makes the sustainability a key feature of smart urban spaces. According to these authors, the solutions provided by the ICT market are one of the ways that sustainable cities are also smart city:

> In these settings, an improvement in sustainability relies on the deployment of ICT systems, along with the introduction of appropriate policy interventions and urban planning. (Neirotti et al., 2014, p. 27)

Steenbruggen et al. (2015) points to the list of smart cities with environmental sustainability through rational management of resources, which necessarily demand information about the use of these natural resources and the structure of the cities. The organization and analysis of this information can boost in the responsiveness to urban demands. In this context, the authors said that the term "smart city" should be seen as the sum of investment in social and human capital, mobility, physical and

Table 1.1 Convergent features in approaches to smart cities

	Characteristic	Authors
1	Usage of ICTs	Weiss et al. (2015), Nam and Pardo (2011), Klauser et al. (2014), Mitchell (2007), Citrigno et al. (2014), Al-Nasrawi et al. (2016), and Luque-Ayala and Marvin (2015)
2	Enhancement of Human Capital	Weiss et al. (2015), Nam and Pardo (2011), Klauser et al. (2014), Citrigno et al. (2014), Al-Nasrawi et al. (2016), Luque-Ayala and Marvin (2015), and Steenbruggen et al. (2015)
3	Institutional Relations	Weiss et al. (2015), Nam and Pardo (2011), Klauser et al. (2014), and Steenbruggen et al. (2015)
4	Government-citizen relations	Weiss et al. (2015), Calzada and Cobo (2015), Al-Nasrawi et al. (2016), and Steenbruggen et al. (2015)

Source: The authors (2016)

digital infrastructure, incentives for proper economy, consideration of natural resources and participatory governance. According to the authors, these elements together should allow an increase in the quality of life.

Analyzing these various studies, it can be seen that, although the concept of smart city has a number of possible settings, the approaches merge to the presence of some essential aspects (Table 1.1): firstly, the use of technology, particularly new ICTs (based on use of Big Data in real time, augmented reality, geolocation, etc.) which work with supporting infrastructure to get information, supply services and even for decision making; secondly, promotion of training and qualification of human capital, especially for the active participation of citizens in urban decisions and actions; thirdly, the *modus operandi* of the institutional relations, should be integrated and encourage the cooperation, reducing bureaucracy and transparency; finally, a fourth common feature of approaches, and that is directly related to the previous three is the relationship government-citizen. Studies tend to emphasize that an important aspect of smart cities is the need to increase the capacity of communication between government and citizens, to send and receive information in a quick process that encourage mutual trust and expand the feedback information.

These four principles will be addressed again in sequence. They form the basis for the correlation methodology proposed in this study.

1.4 Resilient Cities

Resilience is a concept used in different areas of knowledge, with wide application in ecologic studies and, more recently, applied in social sciences (Suassuna, 2014). Over time, different conceptual approaches were emerging, especially those related to environmental perspective and a marketing perspective.

The most widespread concept of resilience, associated to ecological and environmental areas, relates to the ability to absorb an environment disturbance before or after it occurs, with recovery and stability (Adger, 2000). The ecological resilience can also be related to resilience and stability after specifically human changes. Thus, the concept is directly associated with, among other factors, the levels and types of resources and products consumption from people (Cenci & Burmann, 2013).

The "marketing" resilience, in turn, is related to maintaining production from the attractiveness of resources and the viability (Pacheco, 2012). In "marketing-business" resilience, in addition to environmental aspects it is also considered economic growth, social equity and organizational culture. Therefore, it is a process directly linked to decision-making and management (Carvalho, 2014).

In the cities, resilience – called "urban" resilience – is associated with speed stabilization and normalcy after a critical situation (Adger, 2000). Studies on urban resilience allow knowing the mode of action for various social groups in crisis situations and adversity, with the identity of the group and its internal cohesion (Alves, 2015).

These approaches, with nomenclatures and conceptual divisions, allow the concept to be understood from different points of view. In the studies on urban planning, however, all these considerations about resilience converge to the same goal, which is the ability to anticipate and adapt to a critical urban situations.

In this sense, Godschalk (2003) proposes that resilient cities have five basic principles: firstly, the disaster management policy seeks to identify and prevent critical situations; secondly, the research to support the policies management; the thirdly, investments in education programs, which enable population to the face of adversity; the fourth principle is cooperation between institutional sectors, allowing discussion on the subject in various fields of knowledge and action; finally, the proactive attitude, which is creativity value to meet situations.

The Rockefeller Foundation (2014) also proposes that cities should seek a model for resilience (Fig. 1.1), based on actions in four urban dimensions. The first dimension is the strategic management, related with the factors (1.1) efficiency, (1.2) empowerment of actors and (1.3) integrated planning. The second dimension is the well-being related to the factors (2.1) minimization of urban vulnerability, (2.2) diversity and employment and (2.3) effective protection of life. The third dimension is infrastructure and ecosystems, related to (3.1) mobility and communications, (3.2) provision of services and (3.3) reduction of exposure and fragility. Finally, the fourth dimension is economy and society, related to (4.1) factors sustainable economy, (4.2) comprehensive security and (4.3) collective identity and community support.

According to the Rockefeller Foundation (2015), there are seven features which may make a more resilient city (see traces in gray, in the Fig. 1.1), which must be attached to these four acting dimensions:

(a) Reflectivity: constant reinvention capacity as needed, without the use of unique and closed in themselves solutions.
(b) Robustness: physical assets, well-designed and well-managed infrastructure with predictability (anticipation of possible failures);
(c) Redundancy: Ability to have several ways to meet the same demand, with constant and deliberate replacements;

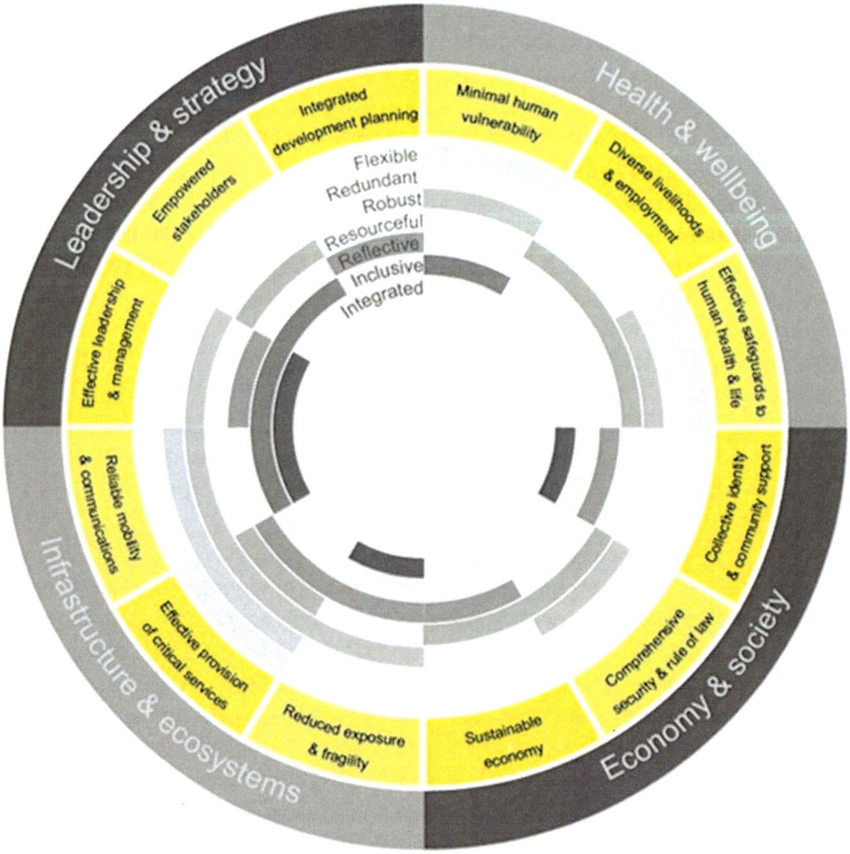

Fig. 1.1 Framework to resilient cities. (Source: The Rockfeller Foundation and ARUP 2014)

(d) Flexibility: Ability to adapt to changes and circumstances;
(e) Full Features: Ability to find solutions quickly and restore quickly the functionality of a given system;
(f) Inclusion: Encourage the engagement of various social knowledge and treatment of critical situations, generating the idea of responsible community;
(g) Integration: Alignment of all the systems that make up and/or manage the environment.

Urban resilience is, according to the specific social and cultural conditions Foundation (2014), directly linked to development and urban prosperity, including the marketing point of view (preservation of capital values). According to this institution, resilience is a concept that articulates elements vulnerable and adaptive capacity. The vulnerability of a city is reflected in the rate of exposure to environmental or climate shocks, in terms of incidence, frequency and magnitude. The ability to adapt, in turn, is related to how governance, institutions, technology and planning are articulated to mitigate and/or minimize the vulnerability. According to this resilience concept, vulnerability and adaptive capacity are quantities that are articulated in proportion, as the adaptability increases, resilience increases; as the vulnerability increases, the resilience tends to decrease. In this sense, a Grosvenor publication (2014) points there are five dimensions linked to vulnerability (environmental conditions): (a) climate, (b) environment capacity, (c) resource capacity, (d) infrastructure and (e) community. The same publication points that adaptive capacity is a condition built by five elements (actions and strategies): (a) governance, (b) institutions, (c) technical and learning, (d) planning systems, and (e) funding structures.

According to the institution, an intersection of several data measuring vulnerability and adaptive capacity has resulted in a ranking of 100 cities more resilient. In this study, it was shown that among the first fifteen placed there are three Canadian cities (Toronto, Vancouver and Calgary, the first two in the best position in this ranking), three European cities (Stockholm, Zurich and Amsterdam), one Australian city (Melbourne) and eight American cities (Chicago, Pittsburgh, Boston, Washington DC, Atlanta, Seattle, New York and Detroit).

Adger (2000) highlights that there is a social group that is the most vulnerable to moments of tension and risk; this group is generally associated with economic issues (that puts it in insecurity). Morrow (2008) supports this argument by arguing that there are specific social and cultural conditions that enhance the intensity of risks and vulnerabilities. The author affirms that a resilient city, therefore, goes beyond the control of risks and should also consider economic, cultural, political and social issues.

The authors, Coaffe et al. (2009) also follow this line of thought, using as an example the terrorist attacks of September 11 to demonstrate that the vulnerability and risks were also present in the social, economic and political dimensions, in addition to terrorism itself. The authors also highlight the importance of using technologies for urban safety and the treatment of vulnerability (p. 253): "The normalized notion of security and resilience has begun to pervade everyday life posing critical questions regarding the relationship between resilience broader policy [...]".

In this sense, Reginaldo et al. (2013) state that the community need to have coping conditions of critical situations, and for this, the communication infrastructure is to assist in making the most independent citizens and more participatory governance (through social networks, for example). Alves (2015) affirms that the use of ICTs – especially those aimed at gathering information in real time – helps planners in decision-making from the increased capacity to assess, control and simulation of actions on the territory.

In this context, we can highlight the participation of ICTs as a possible alternative to mitigation some critical urban situations. This relationship occurs in the sense that the ICTs allow obtaining data in real time, helping the management to know the information and to make decisions.

Folke (2006) calls this process "resilience engineering". Resilience engineering is to build a database as fully as possible so that you can create a kind of "collective memory" (with images, statistics, risk areas, population characteristics, etc.). In this process, it becomes possible to develop new technologies based on existing, that help prevent or mitigate future vulnerabilities.

As the various approaches to smart cities enable the identification the key issues for the most authors, it is also possible to establish these convergences to the urban resilience concept. In this case, the several approaches presented enable the identification the ten points considered essential by the authors to reduce vulnerability and increase adaptive capacity (Table 1.2).

These ten characteristics will be resuming in the next section, to form the basis to the framework of conceptual joints, as proposed in the methodology section.

1.5 Analysis: Conceptual Joints

Sections 1.3 and 1.4 presented the discussion about the several concepts of smart cities and resilient cities, respectively. At the end of each other of these sections, we tried to establish a confluence of concepts, to identify the most relevant and present characteristics of each other.

In this process, smart city concept was summarized in four main features: use of ICT, human capital, institutional relations and management-citizen relations. Urban resilience concept, in turn, was understood in ten basic principles presented in Table 1.2.

Table 1.2 Convergent features in the approaches to urban resilience

	Characteristic	Authors
1	Absorption capacity of a disturb	Adger (2000) and Grosvenor (2014)
2	Production, resources, attractive and viability	Pacheco (2012) and Cenci and Burmann (2013)
3	Stabilization Speed	Adger (2000), Rockfeller Foundation (2014), and Grosvenor (2014)
4	Disaster Management Policies	Godschalk (2003)
5	Training the population to confront situations	Godschalk (2003), Reginaldo et al. (2013), and Alves (2015)
6	Integration between sector	Godschalk (2003), Rockfeller Foundation (2014), Grosvenor (2014), and Carvalho (2014)
7	Constant Reinvention Capacity	Rockfeller Foundation (2014)
8	Infrastructure Robustness	Rockfeller Foundation (2014), Coaffe et al. (2009), Alves (2015), and Folke (2006)
9	Redundancy of predictability	Rockfeller Foundation (2014), Alves (2015), and Folke (2006)
10	Flexibility	Rockfeller Foundation (2014)

Source: The authors (2016)

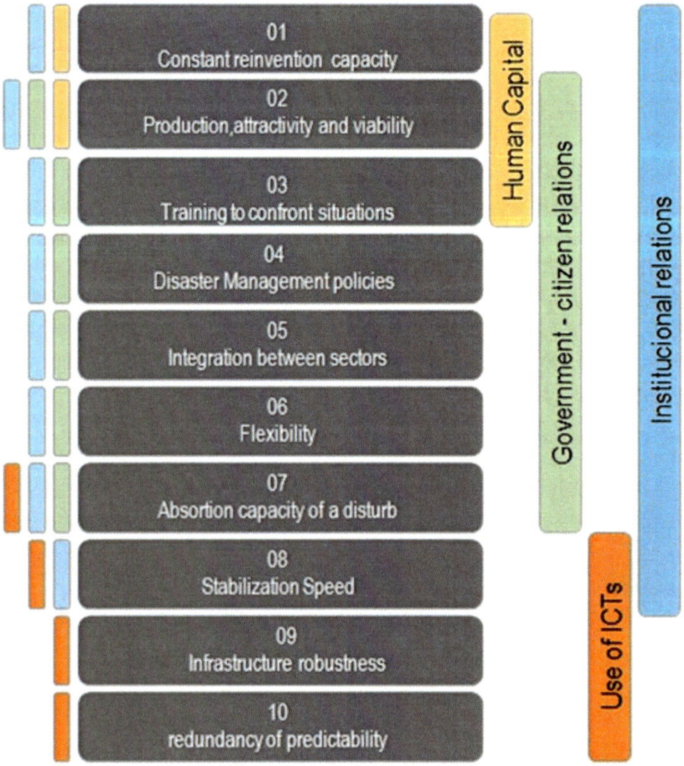

Fig. 1.2 Conceptual joints between smart city and resilient city. (Source: The authors, 2016)

Figure 1.2 shows the conceptual junction proposed here, in order to lay visually how each of the features of a concept are articulated with others concept.

Firstly, it can be noted, initially, that the use of ICT is related, especially, with the support physical infrastructure required for creating urban resilience. This process takes place with the optimization of their own infrastructure and reducing bureaucracy with redundancy of predictability, from the increased use of tools and data. Such data, pass through the mining process, are transformed into information on the territory, helping to give greater speed in responsiveness and stability, and contributing to the disorder be more rapidly absorbed.

Secondly, it is observed that the human capital factor is directly linked to the resilience characteristics that depend on people action: (a) to prepare for the critical situations, (b) encourage cooperation, (c) subsequent reinvention capacity, so that the disturbances are mitigated over time, and (d) the maintenance of production, attractiveness and viability of resources that require, in addition with institutional subsidy, a very significant community effort.

Thirdly, the relationship between urban management and citizens are reflected in coping and managing critical situations policies as well as flexibility and ability to absorb necessary in these times. This relationship is also evident in the need for integration between various sectors, the use of new tools (generating more speed in two-way communication) and consequent stimulus to citizen empowerment in city management processes. In this sense, promoting the exchange of information helps to increase transparency and trust between those who manage and those who inhabit the place, creating resilience in dealing with problems. This is what Calzada and Cobo (2015) determine as articulation between what is individual and what is collective, making it more interesting and more participatory environment in the eyes of citizens. All these elements together, show the relationship between resilience and intelligence helping to avoid the common problem of the smart city concept (highlighted by Kitchin et al., 2016): tendency to technocracy, corporatism and impartiality.

Another aspect of this form of relationship is the obvious need to reduce bureaucracy; this reduction is defended in the logic of smart cities. In this case, the integration depends on institutional management structures less bureaucratic, less amount of paper and documents generated to make any public service request. An alternative to this reduction of bureaucracy is precisely the use of ICTs through digital tools drive and actions monitoring of public administration.

Finally, the institutional relations should be especially considered in all aspects that depend more significantly from management. The integration between various sectors that make up the urban management (services, monitoring, mobility, housing, security, education, etc.) is essential for the basic conditions from resilience will be met.

In this sense, we can highlight two points of conceptual support of smart cities to resilience. First, for better integration of actors and a better distribution of functions (each other knowing what, why and how to do), you should increase the use of performance protocols, detailed for each type of action. Second, the quality of leadership is critical in this process, Nam and Pardo (2011) determine how success factor for these urban projects. These two factors will be reflected in the improvement of the municipal management process and urban resilience.

Finally, also stands out the need to improve lessons learned from previous situations. This is also an element of resilience conceptual that convergence between urban and urban intelligence. The process occurs after a critical situation, which must be analyzed in various optical (actors relations, obtained information, necessary information, made actions, speed response, etc.). All these elements must generate a constant process of evaluation and reevaluation, to integrate the system with successful lessons and failures to be mitigated in the future.

1.6 Conclusions

The debate on urban resilience, despite being recurring issue in academic, tends to receive greater prominence in society only when disorders occur in large-scale affecting directly and severely urban areas.

The discussion on smart cities, in turn, tends to stand – outside the academic researches-, technology and the high financial investment, at disregarding often, human and social aspects involved in this process.

This research sought to discuss how the different conceptual approaches to smart city contribute to enrich the debate on the improvement of urban resilience in observation to studies on urban planning. We try to understand the logic of both processes, identifying convergences and alternative application of the concepts.

It was observed that this link exists and that although academically the concepts can be worked under different approaches, the joint consideration is important and can contribute to a more solidified, efficient, and participatory urban environments.

One of the main points discussed is the possibility of using intelligent processes (via ICT, data dissemination, sharing networks) as an alternative to increase urban resilience. Thus, one can merge the high investment in works with solutions that, as the case may be simpler and cheaper. This occurs even in the sense that, with the use of ICTs, it is possible to recognize the largest territorial vulnerabilities, and from this, make decisions about where, how and why carry out urban transformations.

The transformations in the city should be structured, with adequate speed to every reality with feasible costs to the site, related to long-term scenario and in order to always make the actor citizen involved in the process and encouraged to cooperate.

During this work, it was also presented that the concept of smart city can often mask a process of marketing and "sale" of products related – largely – to the mitigation of urban vulnerabilities.

This research is limited to discussing resilience and urban intelligence in a conceptual way, and although it considers academic and institutional contributions it cannot empirically establish the application of these concepts. We also sought to highlight that both concepts have many diverse approaches (academic, institutional and market), and therefore the attempted convergence is not – by itself – sufficient for knowledge of all possible conceptual joints.

While this research sought to discuss only one of these many possibilities, an interesting debate rises that the combination of these factors can be considered as a form of alert. It was shown that urban intelligence and urban resilience are conceptually converging in many ways. In this sense, one must question if urban resilience can, like urban intelligence, also be sold as a "ready" product and a problem solver, rather than a process of solidification and urban improvement. This discussion is important in order to demonstrate that urban vulnerabilities generate impacts beyond those related to human life and the financial aspects.

In this sense it is suggested, as future studies, the inclusion of other elements that would help enlarging the discussion, such as statistical data, application examples and observation of successes and existing projects failures.

References

Adger, W. N. (2000). Social and ecological resilience: Are they related? *Progress in Human Geography, 24*(3), 347–364. Recovered on March, 2016, from https://groups.nceas.ucsb.edu/sustainability-science/2010%20weekly-sessions/session-102013-11.01.2010-emergent-properties-of-coupled-human-environment-systems/supplemental-readings-from-cambridge-students/Adger_2000_Social_ecological_resilience.pdf

Al-Nasrawi, S., Adams, C., & El-Zaart, A. (2016). A conceptual multidimensional model for assessing smart sustainable cities. *Revista de Gestão da Tecnologia e Sistemas de Informação, 12*(3), 541–558.

Alves, C. M. D. S. (2015). *Urbanismo Participativo e resiliência das comunidades: Especificação de uma aplicação* (Dissertação de mestrado). Programa de pós graduação em Engenharia e Gestão de Sistemas de Informação, Universidade do Minho. Recovered on October, 2016, from http://repositorium.sdum.uminho.pt/handle/1822/40288

Calzada, I., & Cobo, C. (2015). Unplugging: Deconstructing the smart city. *Journal of Urban Technology, 22*(1), 23–43.

Carvalho, C. P. D. (2014). Resiliência empresarial: Uma necessidade ante a sustentabilidade das organizações. *Cambiassu – Revista Científica do Departamento de Comunicação Social da Universidade Federal do Maranhão, 19*(14), 71–92. Recovered on October, 2016, from http://www.periodicoseletronicos.ufma.br/index.php/cambiassu/article/view/3515/1535

Cenci, D. R.; & Burmann, T. K. (2013). Direitos Humanos, Sustentabilidade Ambiental, Consumo e Cidadania. *Revista Direitos Humanos e Democracia, 1*(2), 131–157. Recovered on October, 2016, from https://revistas.unijui.edu.br/index.php/direitoshumanosedemocracia/article/view/338/2460

Citrigno, S., Graziano, S., & Sacca, D. (2014). *Cooperation of smart objects and urban operators for smart city applications.* Recovered on February, 2016, from http://ceur-ws.org/Vol-1156/paper4.pdf

Coaffe, J., Murakami Wood, D., & Rogers, P. (2009). *The everyday resilience of the city: How cities respond to terrorism and disaster.* Palgrave Macmillan.

Folke, C. (2006). Resilience: The emergence of a perspective for social–ecological systems analyses. *Global Environmental Change, 16*, 253–267.

Godschalk, D. R. (2003). Urban hazard mitigation: Creating resilient cities. *Natural Hazards Review, 4*(3), 136–143. Recovered on March, 2016, from http://www.tc.umn.edu/~blume013/Godschalk_urb_haz_mit2003.pdf

Gressler, L. A. (2004). *Introdução à pesquisa: projetos e relatórios* (2nd ed.). Loyola.

Grosvenor. (2014). *Resilient cities: A Grosvenor Research Report.* Recovered on October, 2016, from http://www.grosvenor.com/getattachment/194bb2f9-d778-4701-a0ed-5cb451044ab1/ResilientCitiesResearchReport.pdf

Hall, R. E. (2000). The vision of a smart city. In *Proceedings of the 2nd international life extension technology workshop, Paris.* Recovered on September, 2015, from http://www.osti.gov/bridge/servlets/purl/773961-oyxp82/webviewable/773961.pdf

Holling, C. S., Schindler, D. W., Walker, B. W., & Roughgarden, J. (1995). Biodiversity in the functioning of ecosystems: An ecological synthesis. In C. Perrings, K. G. Mäler, C. Folke, C. S. Holling, & B. O. Jansson (Eds.), *Biodiversity loss: Economic and ecological issues* (pp. 44–83). Cambridge University Press.

Kitchin, R. (2014). From a single line of code to an entire city: Reframing thinking on code and the city. *Code and the City workshop*, 3rd–4th September 2014, Programmable City, NIRSA, National University of Ireland Maynooth.

Kitchin, R., Mc Farlane, C., & Marvin, S. (2016). Introduction. In A. Luque-Ayala, C. M. Farlane, & S. Marvin (Eds.), *Smart urbanism: Utopian vision or false dawn?* (pp. 1–15). Routledge.

Klauser, F., Paasche, T., & Söderström, O. (2014). Michel Foucault and the smart city: Power dynamics inherent in contemporary governing through code. *Environment and Planning D: Society and Space, 32*, 869–885.

Lazaroiu, G. C., & Roscia, M. (2012). Definition methodology for the smart cities model. *Energy, 47*, 326–332.

Luque-Ayala, A., & Marvin, S. (2015). Developing a critical understanding of smart urbanism? *Urban Studies, 52*(12), 2105–2116.

Mitchell, W. J. (2007). Ciudades inteligentes. *UOC Papers Revista sobre la sociedad del conocimiento*, 5. Recovered on June, 2014, from de http://www.redalyc.org/articulo.oa?id=79000502

Morrow, B. H. (2008). *Community resilience. A social justice perspective* (CARRI Research Report, 4).

Nam, T., & Pardo, T. A. (2011). Smart city as urban innovation: Focusing on management, policy, and context. In *Proceedings of the 5th international conference on theory and practice of electronic governance – ICEGOV '11*. pp. 185–194.

Neirotti, P., De Marco, A., Cagliano, A. C., Mangano, G., & Scorrano, F. (2014). Current trends in Smart City initiatives: Some stylised facts. *Cities, 38*, 25–36.

Pacheco, S. M. M. (2012). Resiliência urbana e comercial em áreas centrais. *Geo UERJ, 2*(3), 459–478. Recovered on March, 2016, from http://www.e-publicacoes.uerj.br/index.php/geouerj/article/view/4814/5419

Polit, D. F., & Beck, C. T. (2011). *Fundamentos de pesquisa em enfermagem* (7th ed.). Artmed.

Reginaldo, T., Pacheco, D. C., Baldessar, M. J., & Franzoni, A. M. D. (2013). Redução de riscos de desastres na prática: a participação da Defesa Civil e do município em plataformas digitais de redes sociais. *Revista Democracia Digital e Governo Eletrônico*, 2(9), 66–87. Recovered on October, 2016, from http://www.buscalegis.ccj.ufsc.br/revistas/index.php/observatoriodoegov/article/view/34282/33158

Souza, K. D., & Flanery, T. H. (2013). Designing, planning, and managing resilient cities: A conceptual framework. *Cities, 35*, 89–99. Recovered on March, 2016, from https://www.researchgate.net/publication/259122914_Designing_planning_and_managing_resilient_cities_A_conceptual_framework

Steenbruggen, J., Tranos, E., & Nijkamp, P. (2015). Data from mobile phone operators: A tool for smarter cities? *Telecommunications Policy, 39*(3–4), 335–346. Recovered on September, 2015, from https://doi.org/10.1016/j.telpol.2014.04.001

Suassuna, C. C. A. (2014). *Cidade resiliente: sistema de indicadores dos aspectos institucionais* (Tese de doutorado). Programa de Desenvolvimento Urbano da Universidade Federal de Pernambuco. UFPE.

The Rockfeller Foundation. (2014). *Resilient city framework: City resilience index*. ARUP. Recovered on March, 2016, from https://www.massport.com/media/266329/2014-April_City-Resilience-Framework.pdf

Vanolo, A. (2014). Smartmentality: The smart city as disciplinary strategy. *Urban Studies, 51*(5), 883–898.

Weiss, M. C., Bernardes, R. C., & Cansoni, F. L. (2015). Cidades inteligentes como nova prática para o gerenciamento dos serviços e infraestruturas urbanos: a experiência da cidade de Porto Alegre. *URBE – Revista Brasileira de Gestão Urbana, 7*(3), 310–324.

Chapter 2
Framing 'Resilient Cities': System Versus Community Focused Interpretations of Urban Climate Resilience

Arjan Wardekker

Abstract Building urban resilience to climate change and other challenges will be essential for maintaining thriving cities into the future. Resilience has become very popular in both research on and practice of climate adaptation. However, people have different interpretations of what it means: what resilience-building contributes to, what the problems, causes and solutions are, and what trade-offs, side-effects and other normative choices are acceptable. These different ways of 'framing' climate resilience are hidden in the positive, but sometimes fairly vague, language used to promote it.

Analysis of the framing of 'urban resilience' can distinguish important contrasting preferences regarding the 'most appropriate' way to build urban resilience. This chapter explores two important frames of urban resilience: the 'system resilience' frame, focusing on maintaining urban functions and processes, and the 'community resilience' frame, emphasising urban life, community bonds and self-sufficiency.

The frames used by scientists, policymakers, and stakeholders reflect social uncertainties in climate adaptation, related to values, preferences, and goals. They entail different visions on the urban future, leading to different potential realisations of climate change adaptation. Leaving them implicit can result in a 'dialogue of the deaf', potentially leading to adaptation failure.

Urban decision-makers and stakeholders will need to investigate and develop a clear vision on what they mean by urban resilience: what are the goals, and who's or what's resilience are we talking about? Explicit exploration of the current and potential frames will help to cultivate meaningful discussion on the choices and trade-offs to be made in developing climate-resilient urban futures.

A. Wardekker (✉)
Centre for the Study of the Sciences and the Humanities, University of Bergen, Bergen, Norway
e-mail: Arjan.Wardekker@uib.no

O. F. González Castillo et al. (eds.), *Urban Resilience: Methodologies, Tools and Evaluation*, Resilient Cities, https://doi.org/10.1007/978-3-031-07586-5_2

Keywords Framing analysis · Urban resilience · Climate change adaptation · Climate governance

2.1 Introduction

The majority of the world's population is currently living in cities and the urban population is expected to increase from 3.9 billion in 2014 to 6.4 billion in 2050, rising from 54% to 66% of the total population (UN, 2014). In developed countries, the percentages will be even higher. For example, in Europe, the percentage of the population living in urban areas will rise from 73% to 80% (UN, 2014). The world is – and its future will be – increasingly urban.

At the same time, cities face numerous challenges. While we've come a long way from the days of the Industrial Revolution, with its polluted air and water, poor living and working conditions, and disease, cities are now increasingly faced by more pervasive issues that play over long periods of time and cannot easily be solved at the local level. Examples include the aging population, refugees, (socio-) economic challenges, transboundary pollution, security risks, emerging technologies, and climate change. Such issues are inherently complex and uncertain, and decision-makers will need to find ways to deal with ignorance and surprise. Less complex threats could in the past often be solved using a 'predict & prevent' approach (cf. Dessai & Van der Sluijs, 2007; Capela Lourenço et al., 2014): one simply studies the risk, figures out the magnitude and chances, and designs and dimensions the correct responses through policy, legislation or engineering accordingly. That approach is not always suitable for dealing with today's complex grand challenges, where simply calculation the risk and selecting the 'best' option is often not possible. In these situations of high uncertainty and potential surprise, one may however still have enough knowledge to find ways to strengthen the resilience of the impacted system (e.g. a city, region, or society in general) (Barnett, 2001; Dessai & Hulme, 2004; Dessai & Van der Sluijs, 2007; Wardekker, 2011; Capela Lourenço et al., 2014; Thissen et al., 2017). In recent years, resilience has indeed become a prominent topic in urban research and policy.

Resilience is a concept that emerged in ecology in the 1960s. It was an explanatory concept, highlighting the various processes in dynamic complex systems that produce the high degree of stability and adaptability that we observe in natural ecosystems in the face of a wide range of external perturbations and abiotic conditions (e.g. Walker et al., 2004; Folke, 2006). It has since been applied by numerous disciplines, ranging from engineering to psychology to disaster risk management. Similarly, the concept has gained much popularity in various policy fields. For example, the OECD (2014) recently indicated resilience as an important aspect of coping with critical risks, and the UN's (2015) Sustainable Development Goals (UN, 2015) and Habitat III New Urban Agenda (UN, 2016) use the term in relation to various specific threats and goals as well.

Walker et al. (2004) define resilience of 'social-ecological systems', which includes cities, as: "the capacity of a system to absorb disturbance and reorganize while undergoing change so as to still retain essentially the same function, structure, identity, and feedbacks". Examples of definitions for urban resilience specifically include: "the capacity of individuals, communities, institutions, businesses, and systems within a city to survive, adapt, and grow no matter what kinds of chronic stresses and acute shocks they experience." (Rockefeller Foundation, 2016), or "the ability of an urban system - and all its constituent socio-ecological and socio-technical networks across temporal and spatial scales - to maintain or rapidly return to desired functions in the face of a disturbance, to adapt to change, and to quickly transform systems that limit current or future adaptive capacity" (Meerow et al., 2016).

2.2 Urban Resilience to Climate Change

Climate change is one of these serious, but complex urban challenges, particularly in low-lying deltas. It is expected to impact cities in a diversity of ways: multiply types of impacts on multiple scales and time-frames, effecting multiple groups of people. Examples of sectors that can be impacted include: water management (e.g. flooding, drought, freshwater supply, sea level rise), critical infrastructures (energy, ICT, transportation), health (heat, air quality, diseases, etc.), tourism (in positive or negative ways, depending on the location), housing and communities (through impacts of various potential disasters and long-term changes), food supply, and urban nature and biodiversity (e.g. IPCC, 2014). These in turn have secondary effects on the urban economy, urban resource dynamics and the population. For urban water management, these could include a rising sea level, changes in precipitation (e.g. intense rain showers in the summer, increased river flood risks in wet periods), and heat & drought (more water needed, but less available for water supply, food production, energy production, etc.).

The effects of climate change are both highly uncertain and interact in complex ways with the systems they impact and numerous other issues and trends, such as increasing populations, shift of populations towards urban areas, and other environmental, economic, technological, and societal trends. Climate change provides a complex and highly interdisciplinary issue that cities and regions will need to adapt to. Enhancing the resilience of cities is needed to make the rapidly urbanizing world less vulnerable to climate change related disturbances and surprises, to enable quick and flexible responses to crises and long term issues, and to maintain a thriving city into the future (e.g. Wardekker et al., 2010).

Resilience has recently gained popularity in the fields of (urban) sustainability and climate change adaptation as well, particularly in relation to flood risk management and urban planning. Key disciplinary subfields involved in urban climate resilience include: urban ecology, urban & regional economics, hazards & disaster risk reduction, and governance & institutions (Leichenko, 2011). Resilient development has become a central concept in IPCC's Fifth Assessment Report (IPCC, 2014). At the local level, the ICLEI Local Governments for Sustainability network has been promoting resilience and organising 'Resilient Cities' congresses since 2010 (ICLEI, 2016), and the 100 Resilient Cities network has been "Helping cities around the world become more resilient to the physical, social, and economic challenges that are a growing part of the 21st century" (Rockefeller Foundation, 2016), for instance by stimulating the appointment of Chief Resilience Officers in cities and by providing tools and support. These organisations often cite weather and climate related disturbances as one of the key threats to cities. Weather-related disasters and disaster recovery efforts in urbanised regions, such as hurricane Katrina in New Orleans and Sandy in New York, or the 2003 European Heat Wave, further highlights the importance of improving urban climate resilience.

However, both the impacts and the measures taken to make cities more resilient to these changes or otherwise cope with them, can strongly influence e.g. the aesthetical, spatial and structural setup of cities, neighbourhoods and buildings, urban economics and business models and interests of various companies and other stakeholders, and the daily lives of citizens. Most would agree that it is important to build resilience to climate change. However, considering the above, it should not be surprising that different stakeholders and citizens will have a different take on how cities should go about building urban climate resilience.

2.3 Framing Climate Change Adaptation and Urban Climate Resilience: Resilient Future of Who or What, Exactly?

The notion of 'strengthening urban resilience' provides a distinctly positive way to discuss urban climate adaptation and other urban policy agenda's (McEvoy et al., 2013). It is also relatively open to interpretation and tailoring; in itself it does not prescribe any specific way to measure or evaluate resilience or specific type of policy options that would enhance resilience. This has both advantages and disadvantages. On the one hand, it helps to bring together and inspire a wide variety of stakeholders with diverse interests and goals, and allows decision-makers to tailor the implementation of urban resilience to specific local problems and requirements. In that respect, the vagueness of the term 'resilience' helps it to function as a 'boundary object', connecting the many fields, sectors and stakeholders involved in the urban system (Brand & Jax, 2007; Meerow et al., 2016). This can trigger bottom-up innovation. On the other hand, stakeholders run a serious risk of 'talking past each other,' as they may have very different ideas on what urban resilience really means in practical terms, and how their respective interests fit into that picture. These differences can remain hidden in the discussions, until the moment arrives to make the matter more concrete: when actual interventions or evaluation criteria need to be designed.

The conceptual malleability of urban resilience means that different people will paint different pictures of what a 'resilient urban future' will look like. In other words, people will have different ways of *framing* urban resilience. Framing means that people "select some aspects of a perceived reality and make them more salient... in such a way as to promote a particular *problem definition*, *causal interpretation*, *moral evaluation*, and/or *treatment recommendation* for the item described" (Entman, 1993). Often, this happens subconsciously, in a taken-for-granted way. People can differ in their framing of, for instance:

- What is the problem, really?
- Who's problem is it? (e.g. who's responsible for causing, exacerbating and solving it; who should have a say, and in which way, in making decisions about it?)
- What are the most important causes of this problem?
- What important values are being threatened?
- What should be done about it, and by whom?

This is not simply a matter of semantics. Diverging interpretations of what urban resilience means, also imply differences in preferences regarding how resilience can best be achieved (what is appropriate, effective, efficient, etc.), with which options and interventions, and how it can be evaluated using what metrics and tools. They highlight specific problem aspects and can strongly colour what people perceive as 'valid', 'sensible' policy options and (spatial, structural, social, or other) interventions, fair distributions of burdens, appropriate governance arrangements,

Table 2.1 Example of different frames of climate change adaptation in general, grouped into four strategic perceptual contrasts

	Goal orientation and focus	
Perceptual distance	Promotion orientation	Prevention orientation
Distal view (long-term, broad perspective)	*Social progress frame* Defines the issue as improving quality of life or harmony with nature *Middle way frame* Puts the emphasis on finding a possible compromise position between polarized views Example: Plan to reconcile adaptation and mitigation	*Morality/ethics frame* Defines the issue in terms of right or wrong; respecting or crossing limits *Pandora's box frame* Defines the issue as a call for precaution in face of possible impacts or catastrophe Example: Al Gore's movie, An inconvenient truth
Proximal view (short-term, narrow perspective)	*Economic development frame* Defines the issue as investment that improves competitiveness *Conflict/strategy frame* Defines the issue as a game among elites, a battle of personalities or groups Example: Climate Proof City	*Scientific uncertainty frame* Defines the issue as a matter of what is known versus unknown *Public accountability frame* Defines the issue as responsible use or abuse of science in decision-making Example: Sea level discussion

Source: Wardekker et al. (2009), De Boer et al. (2010); adapted from Nisbet (2009)

and relevant scientific and policy information and tools for decision-making, while obscuring others (De Boer et al., 2010; Wardekker et al., 2009). See Table 2.1 for an example of different frames of climate change adaptation in general. De Boer et al. distinguish between frames that are either focused on prevention or promotion, and distal (broad) or proximal (narrow). An 'economic development frame' of adaptation, for example, would make different impacts, options, actors, information, values and choices relevant than a 'morality/ethics' frame would. Similarly, Fünfgeld & McEvoy (2010) and McEvoy et al. (2013) distinguish framings that are focused on either broader notions of hazards, climate impacts specifically, risk management, or vulnerability.

Different frames entail different goals, boundary conditions, and trade-offs. In that respect, frames can be considered as a form of uncertainty, specifically social uncertainty, in climate change adaptation. They involve different assumptions underlying specific paths that resilient urban adaptation could (when viewed 'from the outside') our should (from the perspective of a specific urban actor) take. Depending on what actors are more powerful or persuasive and what frames become most dominant, actual adaptation pathways can go in different directions. As such, it can be seen as a type of scenario uncertainty (cf. Walker et al. 2003; Mathijssen et al., 2008), particularly related to value-ladenness, although there will be considerable (recognized and unrecognized) ignorance as well regarding how varied local actors frame and interpret resilience for the local situation more specifically.

Considering the above, we will need to be explicit (or at least explore our differences) in designing and building resilient urban futures: The resilient futures of who or what, exactly? What are our goals in building resilience, and who can contribute in what way?

2.4 Frames of Urban Resilience in the Scientific Literature

Before exploring the framing of urban resilience within the resilience literature, it is good to note that the move that cities are making towards urban climate resilience, is in itself a way of reframing the debate on climate change adaptation (McEvoy et al., 2013). Rather than placing the focus on preventing climate change impacts and keeping out the threat, it shifts the narrative to one describing how to develop a 'good' city. In the terms of De Boer et al. (2010) (see Table 2.1), it reframes urban adaptation from prevention to promotion. Whether urban resilience is, in those terms, proximal or distal, will vary from case to case, and from time to time. E.g., the work in Rotterdam on urban resilience first focused fairly narrowly on specific neighbourhoods in the city that are not protected by dikes and on the economic competitiveness of the city (Wardekker et al., 2010), but in the recent release of its Resilience Strategy (Municipality of Rotterdam, 2016), the focus has broadened towards one of urban future proofing on multiple topics and goals. Cities also differ in whether they focus more on short- or long-term aspects of urban resilience building and of climate change itself.

While there are presently only a handful of papers that explore the framing of (urban) resilience, that aspect of short-term versus long-term resilience is one that received particular attention (e.g. Davoudi & Porter, 2012; Davoudi, et al., 2013; Sakai & Dessai, 2015; Meerow et al., 2016). Short-term resilience is mainly focused on absorbing shocks and a quick return to equilibrium: maintaining the status quo. This relates to the classic notion of 'engineering' resilience (Folke, 2006), which lends itself well to relatively narrow (proximal) interpretations of the problems, causes, and solutions related to climate change adaptation (cf. De Boer et al., 2010). Long-term resilience deals with the inevitability of change in complex, dynamic systems, with transformation, adaptability & flexibility and co-evolving with trends. It relates to the classic notion of 'socio-ecological' resilience (Folke, 2006), which lends itself to relatively broad (distal) interpretations of problems, causes, and solutions (cf. De Boer et al., 2010). Davoudi et al. (2013) coin the term 'evolutionary' resilience, and 'dynamic' resilience might also be a good description.

In exploring other potential frames of urban resilience, it is useful to reflect on the key choices that need to be made, or are implicitly made, in resilience thinking. Meerow et al. (2016) argue that in developing urban resilience, decision-makers will need to explicitly reflect on five questions: resilience of who, what, when, where, and why, exactly? These lead to different choices, priorities, and trade-offs. Similarly, Chelleri et al. (2015) observe that there are key trade-offs to be made in urban resilience thinking regarding temporal scales and spatial scales. The when/temporal scale dimension is reflected in the literature discussed above on the short versus long term frames. The other dimensions have so far not received much attention.

Below, I will explore another potential set of frames that relate to Meerow's who, what and why, and several of Chelleri's spatial trade-offs: whether the focus of urban resilience building is on urban systems or urban communities. Both seem to relate to different disciplinary origins and interpretations of resilience. The System Resilience frame perceives resilience from the city level and how urban systems functions. This notion relates to the origins of resilience in the system dynamics and ecological literatures. The Community Resilience frame perceives resilience from the level of communities and individuals, and how their relations and abilities help them cope with adversity. This notion relates much more to the strands of resilience thinking that developed in the psychological literature. Both seem 'natural' ways of thinking about resilience, but they result in different priorities and perceptions regarding the problems, causes, values, and solutions. They may have different notions of what mechanism lead to resilience ('resilience principles'), what information is particularly relevant, and who should be 'in the driver's seat' when building urban resilience to climate change. A summary is displayed in Table 2.2. In the following two paragraphs, I will further explore the System Resilience and Community Resilience frames.

Table 2.2 Comparing the System Resilience and Community Resilience frames for urban resilience: framing problems, causes, morals, and remedies

	Framing:	
What's the:	System resilience	Community resilience
Problem	Threat to functioning of urban system	Threat to urban life and social cohesion
Causes	Disruption of resource flows and activities	Societal disruption, hampering of daily life
Moral judgements	Some subsystems and infrastructures may be prioritised (e.g. labelled as 'critical' or seen as politically more important), according to the city decision makers' goals	Moral weighing and relative importance of issues such as social equity, public participation, and impacts on and taking care of vulnerable groups
Remedies	Engineer ways into the urban systems to deal with disturbances and/or enhance the various buffers	Improve social support networks, strengthen urban identity, improve people's skills and education

2.5 A System Resilience Frame: Climate Change as a Challenge to Urban Functioning

Resilience thinking developed in a literature that was oriented on systems analysis, particularly ecology and system dynamics. Many approaches to urban resilience are consequently also rooted in this literature. Urban resilience is, e.g. "the ability of the city to maintain the functions that support the well-being of its citizens" (Da Silva et al., 2012), conceptualizing cities as systems with components, functions, and flows of e.g. resources, materials, and people (e.g. Wardekker et al., 2010; Meerow et al., 2016). Typical 'resilience principles' are derived from system dynamics, such as: homeostasis, buffer capacity, system redundancy, interconnectivity and system openness and dynamism (e.g. Watt & Craig, 1986; Wardekker et al., 2010; Shutters et al., 2015; Martin & Sunley, 2015). See e.g. Eraydin & Taşan-Kok (2013), Biggs et al. (2015), Sharifi & Yamagata (2016), and Wilk (2016) for explorations of resilience principles related to System Resilience. This framing of urban resilience is analytically focused and primarily outcome-oriented. Larger stakeholders and authorities are often natural/key players, particularly when the analysis is performed at the city level.

The Problem(s) System resilience is a framing that seems common in discussions on urban resources, infrastructure and services. Climate-related disturbances, such as extreme weather events (short-term shocks) or sea level rise (long-term stresses), threaten the flow of goods or traffic, continued delivery of urban services, and fulfilling urban functions and needs. For example, flooding may close roads, prolonged heat and drought may threaten the cooling water supply of a power plant, and rising water tables may result in areas becoming unsuitable for housing. Salient problems include particularly those that hamper the smooth functioning of the urban system. A potential blind spot is that relatively short or (at city level) minor disturbances could be overlooked, while they may have disproportionate impacts on specific subsystems or (vulnerable) subpopulations.

The Causes Climate change will have large impacts on urban systems that lack the ability to plan/prepare for, absorb, recover, and/or adapt to climate change-related disturbances (and their combined effects with other trends that also influence urban vulnerability or resilience). Such weaknesses may lie in, for instance, the governance structure, for instance if it is inflexible or lacks the ability to look and plan ahead or adapt, spatial planning, physical infrastructure, the specific sources of resources and routes to obtain these, and/or the actors involved. For instance, if electricity is brought into the system via a single power line, that system cannot absorb the impact of a section of this line going down.

Moral Judgements While this framing rarely focuses explicitly on moral and value aspects, these are certainly present. A particularly prominent one is in setting priorities. In assessing the level of impact that disturbances have on specific urban functions, processes, and actors, one will need to establish whether or not that level of impact is acceptable or not. Implementing measures to reduce the impacts will cost money and effort (which could be spent elsewhere), and any option to increase the resilience for one subsystem, sector, neighbourhood or population would decrease the resources available for –and may even physically reduce resilience of– others. Actions have pros and cons, and sometimes unintended consequences. A similar evaluation is whether resilience-improvement should focus on the current population or on future populations/generations.

Suggested Remedies Key adaptation strategies and options could focus on enhancing the capacity of (sub)systems, sectors and actors to absorb and recover from disturbances, e.g. by enhancing redundancy, omnivory, or buffer capacity. Reaction to disturbances could be improved by fast mobilization of resources ('high flux'), feedback mechanisms ('homeostasis'), or by avoiding overly hierarchical or bureaucratic decision-making ('flatness'). Similarly, disaster preparedness can be improved by investing in monitoring, foresight, and local practical knowledge development. Adaptability can be enhanced by increasing flexibility, resourcefulness, and learning capacity. For discussion of such 'resilience principles', see e.g. (Wardekker et al., 2010; Da Silva et al., 2012).

2.6 A Community Resilience Frame: Climate Change as a Challenge to Social Cohesion

Community resilience is "a process linking a network of adaptive capacities (resources with dynamic attributes) to adaptation after a disturbance or adversity" (Norris et al., 2008). It entails "ongoing and developing capacity of the community to account for its vulnerabilities and develop capabilities that aid the community in… (1) preventing, withstanding, and mitigating… (2) recovering… (3) using knowledge from a past response" (Chandra et al., 2010). This framing of urban resilience is fairly process-oriented; the outcomes are important, but the quality and fairness of the process matter much as well. Typical 'resilience principles' are derived from social science literatures, such as: social networks, leadership, engagement, information flow, encouraging skills & learning, societal partnerships, societal equity (e.g. Norris et al., 2008; Chandra et al., 2010; Berkes & Ross, 2013). See e.g. Ronan & Johnston (2005), Norris & Stevens (2007), Norris et al.(2008), Twigg (2009), IFRC (2011), Berkes & Ross (2013), and Brown (2016) for explorations of resilience principles related to Community Resilience. The Community Resilience frame focuses on the way communities are impacted by disturbances, but also places the ball for dealing with these in their court. Citizens and small stakeholders are natural/key players.

The Problem(s) Community resilience seems to be discussed particularly in the context of disaster preparedness and psychology, discussing how disturbances impact a community, the relations in that community, and the ways in which communities deal with and learn from disturbances. Climate change would in this frame lead to disasters and other disturbances that cause societal disruption, e.g. break up social networks or hamper daily urban life, and decrease physical, financial, and mental wellness. Salient problems include particularly those that directly impact communities, such as flooding and health impacts. A potential blind spot is that it may focus primarily on shock-resilience (i.e. disasters; short-term events), and neglect slower, creeping stresses.

The Causes Climate change may have the severest impacts on communities that have weak community ties, that have low adaptive capacities, including low economic development, social capital, information & communication, and community competence (cf. Norris et al., 2008). These can be low if the social bonds are weak, the community has little access to resources (e.g. financial, informational) which are essential for self-sufficiency and adaptability, low education, and/or its adaptive capacity is already hollowed out by other problems.

Moral Judgements A key moral issue in community resilience is the matter of social and resource equity. Often, those populations that are already less well-off or have significant gaps in terms of wealth, education, and/or health, are also the ones that are most vulnerable to disasters and disturbances in general. Another key value in community resilience relates to a perceived right to public access to information & resources, and public involvement in decision-making (and perhaps also in adaptation research and adaptation implementation). If the community is to be the problem owner, it will want a say in how it is analysed and adapted.

Suggested Remedies Key adaptation strategies and options would focus on enhancing communities' capacity to cope with disturbances in a self-sufficient way.

Community refers to citizens, as well as local businesses, NGO's, and policy actors. As a basis, improving basic living conditions, education, health, wellbeing, social support networks, and social participation in general would be helpful. More advanced strategies would focus engaging communities in research, decision-making, and implementation, by providing or helping them develop the tools and resources to do so. Recent trends such as citizen science, city labs, open data, and bottom-up citizen-led adaptation & sustainability initiatives are a key example of this. An important condition, is that some decision power will need to be distributed to the community –at the very least, their efforts will need to have a clear influence in the decision-making process.

2.7 Conclusions

Building urban resilience to climate change and other urban challenges will be essential for maintaining thriving cities into the future. People have different interpretations of what that means, however, and they can frame the challenge of building urban climate resilience in different ways.

Two important frames of urban resilience include the 'System Resilience' frame, which focuses on maintaining functions and processes, and the 'Community Resilience' frame, which emphasises urban life and community capacity & self-sufficiency. Both seem 'natural' ways of thinking about resilience, but they result in different priorities and perceptions regarding the problems, causes, values, and solutions. They have different notions of what mechanism lead to resilience ('resilience principles'), what information is particularly relevant, and who should be 'in the driver's seat' when building urban resilience to climate change. Such frames do not necessarily exclude each other, but they do highlight – and obscure – other important aspects of urban resilience. Since this framing is often done in a subconscious, taken-for-granted way, people may simply ignore other frames: important aspects may be overlooked and key stakeholders may fail to connect and collaborate.

The different frames can lead to different practical realisations of climate change adaptation in cities. The System Resilience favours adaptation options and strategies that can be expressed system-analytically and top-down. This could lead to a focus on resources (water, energy, information, etc.) and similar flows (financial, traffic, etc.), and resilient infrastructure to support these. This links to broader urban policies on infrastructure, economy, ICT, and public utilities. A Community Resilience frame favours options and strategies that are people-centric and bottom-up. This lends itself well to a focus on education, citizen participation, quality of life, enhancing resilience of vulnerable groups, and local/neighbourhood initiatives. This links to broader urban policies on education, welfare, health, and housing. Each has its advantages and pitfalls and each may benefit different groups within the city.

In practice, there will likely be a wider variety of frames among stakeholders, citizens, and decision-makers in different cities, focusing on different details and topics in relation to what the relevant problems, causes, moral judgements, and remedies are for urban climate resilience. To prevent 'dialogues of the deaf', urban decision-makers and stakeholders will need to investigate and develop a clear vision on what they mean by resilient urban futures: what are the goals, and who's or what's resilience are we talking about? That way, stakeholders can learn from each other's framing and explore (and hopefully reduce) the pitfalls within their own. Explicit exploration of the current and potential frames will help to cultivate meaningful discussion on the choices and trade-offs to be made in developing resilient urban futures.

Acknowledgements The author wishes to thanks the Research Council Norway (RCN) for providing travel funding to a workshop and conference, under project no. 246891, SAMKUL UC4A: 'Understanding cultural conditions for climate change adaptation', which helped develop and refine the ideas in this paper.

References

Barnett, J. (2001). Adapting to climate change in Pacific Island countries: The problem of uncertainty. *World Development, 29*, 977–993.

Berkes, F., & Ross, H. (2013). Community resilience: Toward an integrated approach. *Society & Natural Resources, 26*(1), 5–20.

Biggs, R., Schlüter, M., & Schoon, M. L. (2015). *Building principles for resilience: Sustaining ecosystem services in social-ecological systems*. Cambridge University Press.

Brand, F. S., & Jax, K. (2007). Focusing the meaning(s) of resilience: Resilience as a descriptive concept and a boundary object. *Ecology & Society, 12*(1), 23.

Brown, V. (2016). *Community resilience to climate change disasters: Comparing how Rotterdam and New York City approach community resilience in policy*. MSc thesis. Utrecht University.

Capela Lourenço, T., Rovisco, A., Groot, A., Nilsson, C., Füssel, H. M., van Bree, L., & Street, R. B. (Eds.). (2014). *Adapting to an uncertain climate: Lessons from practice*. Springer.

Chandra, A., Acosta, J., Stern, S., Uscher-Pines, L., Williams, M. V., Yeung, D., Garnett, J., & Meredith, L. S. (2010). *Building community resilience to disasters: A way forward to enhance national health security*. RAND Corporation.

Chelleri, L., Waters, J. J., Olazabal, M., & Minucci, G. (2015). Resilience trade-offs: Addressing multiple scales and temporal aspects of urban resilience. *Environment & Urbanization, 27*(1), 181–198.

Da Silva, J., Kernaghan, S., & Luque, A. (2012). A systems approach to meeting the challenges of climate change. *International Journal of Urban Sustainable Development, 4*(2), 125–145.

Davoudi, S., & Porter, L. (Eds.). (2012). Resilience: A bridging concept or a dead end? Reframing resilience: Challenges for planning theory and practice. Interacting traps: Resilience assessment of a pasture management system in Northern Afghanistan. Urban resilience: What does it mean in planning practice? Resilience as a useful concept for climate change adaptation? The politics of resilience for planning: A cautionary note. *Planning Theory & Practice, 13*(2), 299–333.

Davoudi, S., Brooks, E., & Mehmood, A. (2013). Evolutionary resilience and strategies for climate adaptation. *Planning, Practice & Research, 28*(3), 307–322.

De Boer, J., Wardekker, J. A., & van der Sluijs, J. P. (2010). Frame-based guide to situated decision-making on climate change. *Global Environmental Change, 20*(3), 502–510.

Dessai, S., & Hulme, M. (2004). Does climate adaptation policy need probabilities? *Climate Policy, 4*, 107–128.

Dessai, S., & van der Sluijs, J. P. (2007). *Uncertainty and climate change adaptation: A scoping study*. Utrecht University. Available at: http://www.nusap.net/downloads/reports/ucca_scoping_study.pdf

Entman, R. M. (1993). Framing: Toward clarification of a fractured paradigm. *Journal of Communication, 43*(4), 51–58.

Eraydin, A., & Taşan-Kok, T. (Eds.). (2013). *Resilience thinking in urban planning*. Springer.

Folke, C. (2006). Resilience: The emergence of a perspective for social–ecological systems analyses. *Global Environmental Change, 16*, 253–267.

Fünfgeld, H., & McEvoy, D. (2010). *Framing climate change adaptation in policy and practice*. Victorian Centre for Climate Change Adaptation Research. Available at: http://www.vcccar.org.au/sites/default/files/publications/Framing_project_workingpaper1_240611_1.pdf

ICLEI (2016). Resilient city. Website: http://www.iclei.org/activities/our-agendas/resilient-city.html. Accessed 24 Feb 2016.

IFRC. (2011). *Characteristics of a safe and resilient community: Community based disaster risk reduction study*. International Federation of Red Cross and Red Crescent Societies.

IPCC. (2014). *Climate Change 2014: Impacts, adaptation, and vulnerability*. Cambridge University Press.

Leichenko, R. (2011). Climate change and urban resilience. *Current Opinion in Environmental Sustainability, 3*, 164–168.

Martin, R., & Sunley, P. (2015). On the notion of regional economic resilience: Conceptualization and explanation. *Journal of Economic Geography, 15*(1), 1–42.

Mathijssen, J., Petersen, A., Besseling, P., Rahman, A., & Don, H. (2008). *Dealing with uncertainty in policymaking*. CPB, MNP, RAND Europe.

McEvoy, D., Fünfgeld, H., & Bosomworth, K. (2013). Resilience and climate change adaptation: The importance of framing. *Planning Practice & Research, 28*(3), 280–293.

Meerow, S., Newell, J. P., & Stults, M. (2016). Defining urban resilience: A review. *Landscape and Urban Planning, 147*, 38–49.

Municipality of Rotterdam. (2016). *Rotterdam resilience strategy: Ready for the 21st century*. Municipality of Rotterdam & 100 Resilient Cities.

Nisbet, M. C. (2009). Communicating climate change: Why frames matter for public engagement. *Environment, 51*, 12–23.

Norris, F. H., & Stevens, S. P. (2007). Community resilience and the principles of mass trauma intervention. *Psychiatry: Interpersonal and Biological Processes, 70*, 320–328.

Norris, F. H., Stevens, S. P., Pfefferbaum, B., Wyche, K. F., & Pfefferbaum, R. L. (2008). Community resilience as a metaphor, theory, set of capacities, and strategy for disaster readiness. *American Journal of Community Psychology, 41*, 127–150.

OECD. (2014). *Recommendation of the council on governance of critical risks*. Organisation for Economic Co-operation and Development.

Rockefeller Foundation. (2016). *100 Resilient Cities Initiative*. Website: http://www.100resilientcities.org. Accessed 24 Feb 2016.

Ronan, K. R., & Johnston, D. M. (2005). *Promoting community resilience in disasters: The role for schools, youth, and families*. Springer.

Sakai, P., & Dessai, S. (2015). *Can resilience framing enable adaptation to a changing climate? Insights from the UK water sector* (SRI Papers, no. 88). Sustainability Research Institute (SRI), University of Leeds, Leeds. Available at: http://www.icad.leeds.ac.uk/WorkingPapers/ICADWorkingPaperNo9_SRIPs-88.pdf

Sharifi, A., & Yamagata, Y. (2016). Principles and criteria for assessing urban energy resilience: A literature review. *Renewable and Sustainable Energy Reviews, 60*, 1654–1677.

Shutters, S. T., Muneepeerakul, R., & Lobo, J. (2015). Quantifying urban economic resilience through labour force interdependence. *Palgrave Communications, 1*, 15010.

Thissen, W., Kwakkel, J., Mens, M., van der Sluijs, J., Stemberger, S., Wardekker, A., & Wildschut, D. (2017). Dealing with uncertainties in fresh water supply: Experiences in the Netherlands. *Water Resources Management, 31*, 702–725.

Twigg, J. (2009). *Characteristics of a disaster-resilient community: A guidance note*. DFID Disaster Risk Reduction NGO Interagency Group. http://discovery.ucl.ac.uk/1346086/

UN. (2014). *World urbanization prospects: The 2014 revision*. Department of Economic and Social Affairs, Population Division. United Nations.

UN. (2015). *Transforming our world: The 2030 Agenda for Sustainable Development*. United Nations.

UN. (2016). *Habitat III: The new urban agenda*. United Nations.

Walker, W. E., Harremoës, P., Rotmans, J., van der Sluijs, J. P., van Asselt, M. B., Janssen, P., & Krayer von Krauss, M. P. (2003). Defining uncertainty: A conceptual basis for uncertainty management in model-based decision support. *Integrated Assessment, 4*(1), 5–17.

Walker, B., Holling, C. S., Carpenter, S. R., & Kinzig, A. (2004). Resilience, adaptability and transformability in social-ecological systems. *Ecology and Society, 9*(2), 5–13.

Wardekker, J. A. (2011). *Climate change impact assessment and adaptation under uncertainty*. PhD thesis. Utrecht University.

Wardekker, J. A., de Boer, J., Kolkman, M. J., van der Sluijs, J. P., Buchanan, K. S., de Jong, A., & van der Veen, A. (2009). *Tool catalogue frame-based information tools*. Utrecht University.

Wardekker, J. A., de Jong, A., Knoop, J. M., & van der Sluijs, J. P. (2010). Operationalising a resilience approach to adapting an urban delta to uncertain climate changes. *Technological Forecasting and Social Change, 77*(6), 987–998.

Watt, K. E. F., & Craig, P. P. (1986). System stability principles. *Systems Research, 3*, 191–201.

Wilk, B. (2016). *Translating the scientific concepts of resilience into a diagnostic tool for urban climate resilience building*. MSc thesis. Utrecht University.

Chapter 3
Townscape Catalogues Toward Urban Resilience and Sustainability

Lucas Períes, María Cecilia Kesman, and Beatriz Ojeda

Abstract This paper gathers the findings of a series of research projects related to the Townscape Catalogues. Since their start in 2010, these projects have been conducted within the Facultad de Arquitectura [School of Architecture] at Universidad Católica de Córdoba.

This article springs from the need to define *Catálogo de Paisaje Urbano* [Townscape Catalogue] as an innovative instrument that applies a landscape-oriented strategy in the study of the city, develops the proposed methodology and defines the three phases it comprises. Before the conclusions, a series of considerations are noted presenting possible derivations of our proposal toward urban resilience and sustainability. The study of the landscape becomes fertile ground here for the convergence of the contributions of multiple disciplines, where, fittingly, we may propose alternative outlooks and perspectives.

As an open research field, urban resilience may encounter conceptual and methodological foundations in the Townscape Catalogues.

Keywords Landscape · City · Heritage · Development

3.1 Towards Urban Planning Centered on a Landscape-Based Approach

This paper gathers the findings of a series of research projects related to the Townscape Catalogues. The instrument generated here aims at orienting urban planning processes around a landscape-based strategy. The knowledge and capabilities attained in previous studies has resulted in the development of new tools to improve

L. Períes (✉) · M. C. Kesman · B. Ojeda
Universidad Católica de Córdoba – Unit Associated with CONICET, Córdoba, Argentina;
https://www.lucasperies.com

O. F. González Castillo et al. (eds.), *Urban Resilience: Methodologies, Tools and Evaluation*, Resilient Cities, https://doi.org/10.1007/978-3-031-07586-5_3

resource management, analysis and usage in order to produce more effective urban policies of sustainable development.

Since its onset in 2010, research has been conducted within Facultad de Arquitectura [School of Architecture] at Universidad Católica de Córdoba in the form of various projects financed on different grants provided by the latter university and Ministerio de Ciencia y Tecnología [Ministry of Science and Technology] of the province of Córdoba. The different research projects have been endorsed by municipal institutions in Córdoba such as Secretaría de Arquitectura [Department of Architecture], Universidad Libre del Ambiente and Instituto de Planificación Municipal [Institute of Municipal Planning]. Cooperation agreements and agreements of cession of results have been signed with the Town Hall. In 2015, their results were declared of provincial interest by Ministerio de Agua, Ambiente y Servicios Públicos [Ministry of Water, Environment and State-run Services] of the government of the province of Cordoba.

The main goal of this set of projects is to contribute to the understanding of the local reality regarding the urban environment, furthering knowledge around the general theory on landscape planning. This applied research may be put into practice by its transference to the local government sphere. We note the need to strengthen and foster cooperation to find solutions that enhance the conditions of urban living. To that effect, it is necessary to particularly focus on underprivileged and vulnerable populations as a means to boost sustainable urban development. The Townscape Catalogues attempt solutions following this line of thought.

3.1.1 Townscape Catalogues

Emergency in the rapid transformation of cities calls for scientific reflection and action generating innovative strategies and instruments in planning urban landscapes. Understanding the city as an ecosystem (Di Pace & Caride Bartrons, 2004) allows a holistic approach that enables its understanding as a complex system in which all manners of exchanges between its components occur, as is the case in any process of this kind.

Urban design and planning require a system-based approach where natural and cultural factors interact. This perspective demands an interdisciplinary outlook to further the issue of urban development towards more comprehensive strategies beyond disciplinary boundaries – environmental, social or economic, among other realms that affect the quality of living and preservation of goods and services pertaining the environment –, acknowledging the benefits of a system-based approach. The study of the landscape becomes fertile ground here for the convergence of the contributions of multiple disciplines, where, fittingly, we may propose alternative outlooks and perspectives.

Even when traditional studies integrate interdisciplinary perspectives, they have limitations on urban issues regarding their approach to their object of study. Even if encompassing, this approach lacks the vision of a city resident-observer.

The concept of landscape provides a concrete orientation to tackle the urban system globally, constantly and environmentally, integrating intangible, experiential and dynamic aspects in the study of the city.

A Townscape Catalogue is a study of the landscape which condenses multiple information in a document which enables the knowledge of tangible and intangible resources of an area of the city, understanding its historic dimension and outlining its character and value. The goal of this study is to define the intended type of landscape as well as the means to attain it through considerations on the quality of the urban landscape. Potentially, this instrument may offer tools for the analysis, inspection and assessment of urban transformations; the application of territorial policies, and the feedback processes set up between instances of transformation and initiatives of city design and planning.

Thus, our research aims at searching and consolidating a new methodological instrument for urban planning from a landscape-based perspective to detect specific needs for landscape preservation, enhancement and recuperation. The design and execution of a Townscape Catalogue constitutes a novel mechanism compared to previous work that tackles territorial planning on a regional scale through similar strategies. Our research aims at generating our own action-oriented instruments for their application on a local context, scale and reality within the urban sphere.

3.1.2 The Landscape Approach

Reading the city landscape opens interpretations of the physical-spatial reality where man is immersed, creating systems of meaning, hierarchy and values, and, by cultural association, setting up mechanisms in the public imagination to allow us to understand the landscape as imagery (in the broadest sense). As Naselli believes, "landscape might be, then, the result of a dialogue between image and reality. [...] The image is crafted in the intersection between culture and space." (1992, 14). The lack or inadequacy of urban planning threatens the landscape-image to which we refer, and, therefore, endangers cultural-natural heritage and environmental balance, replacing or changing the elements of landscape.

Planning the city should also imply landscape design because planning will unavoidably affect the urban image. Based on this premise, we tackle not only esthetic matters: the ecologic, structural and functional roles of every element added, altered or replaced are key to guarantee the landscape sustainability of the system. Busquets and Cortina define landscape sustainability as the "capacity of a certain landscape to nourish social activities without endangering its essential features and values" (2009, 702).

The procedure proposed in the Catalogue prevents regarding the landscape in an abstract manner – i.e. with cartographic tools – providing information usually overlooked in traditional urban planning activities. Landscape is viewed experientially, holistically and temporally, which allows working on tangible and intangible aspects with special attention to the landscape form and dynamics, identifying its values and shortcomings.

The strategy is based on the study of the image, in agreement with the contemporary theory of the landscape. It derives from Geography and the first studies of our discipline correspond to Lynch (1960) and Cullen (1961). Our key and local reference is César Naselli (1992), who defines landscape as the set of elements and their articulations perceived by an observer. This approach corresponds to the definition of landscape proposed by the European Landscape Convention (2000).

3.1.3 The Proposed Methodology

This methodological strategy has been developed through a continuous research process which encompasses different interconnected projects in which we have worked on specific techniques and materials. On a first stage, in 2010, we studied antecedents in the European context, which we reviewed in the article "Nuevas estrategias de preservación y potenciación del paisaje" [New landscape preservation and enhancement strategies] (Períes et al., 2011). That study led to take *Observatorio del Paisaje de Cataluña* [Catalonia Landscape Observatory] (2005) as a referent, and to adapt their method to urban context (cities) considering that its work tackles the regional scale (shires). On a second stage, we applied this method to the study of a portion of Suquía river in the city of Córdoba, Argentina, in the project "Observatorio del paisaje urbano de la ciudad de Córdoba: hacia un Catálogo del Paisaje del río Suquía, parte 1 centro y pericentro noroeste"[1] [Landscape observatory in the city of Córdoba: Toward a townscape catalogue of Suquía river – Part 1, center and northwest surroundings] (2010–2012). We later carried out a revision and adjustment of the method to overcome the difficulties we encountered when producing the first catalogue. The resulting methodological and operative aspects of the study were presented in the book *Procedimientos para un catálogo del paisaje urbano* [Procedures for a townscape catalogue] (Períes et al., 2013), available on the ISSUU open access digital platform (www.issuu.com). Finally, the method was applied again in the northwest area of Suquía river in the project "Observatorio del paisaje urbano de la ciudad de Córdoba: hacia un catálogo del paisaje del río Suquía, parte 2 suburbano noroeste"[2] [Landscape observatory in the city of Córdoba: Toward

[1]The research team was composed of Lucas Períes, M.Arch. (director); María Cecilia Kesman, M.Arch.; Beatriz Ojeda, M.Arch.; María José Pedrazzani, M.Arch. (researchers responsible); Silvina De Lourdes Barraud, M.Arch; María Noelia Mattio, architect (collaborating researchers).

[2]The research team is made of Lucas Períes PhD. Arch. (director); María Cecilia Kesman, M. Arch.; Beatriz Ojeda, M.Arch.; María José Pedrazzani, M.Arch.; Diana Perazzolo, M. Agronomical engineer (researcher responsible). Collaborating researchers: Silvina De Lourdes Barraud, M.Arch; Natalia Brizuela, architect; Leticia Delacula, agronomical engineer; Cecilia Eynard, biologist; Virginia Fischer, architect; Ana Hofmann, agronomical engineer; Noelia Mattio, architect; María Laura Perasso, biologist; Sofía Urrets Zavalía, agronomical engineer.

a townscape catalogue of Suquía river – Part 2, northwest suburbia] (2013–2016). The results of both application projects were published in the books *Catálogo del paisaje del río Suquía en la ciudad de Córdoba*, Vol. 1 y 2. [Townscape catalogue of Suquía river in the city of Córdoba, Vol. 1 and 2] (Períes et al., 2012, 2016).

It is fitting to note that a process of method design is the focus of interest and guiding principle of our research, as we aim at its possible replicability in other urban contexts. For this reason, we are currently carrying out the research project "Catálogo del paisaje urbano: instrumento innovativo para la planificación urbana con enfoque paisajístico: Parque Sarmiento – Ciudad de Córdoba"[3] [Townscape catalogue: An innovative instrument in urban planning focused on landscape] (2016–2018). This project is directed toward the application of the method in urban spaces with an extended grid structure, those different from linear corridors (rivers, canals, roads or avenues). Its main goal is to produce an efficient, flexible method which is adaptable to the different physical conditions of a city.

3.2 The Design of the Townscape Catalogue

The catalogue design springs from the interpretation of landscape from a territory-based perspective of the area of study which permits the definition of its scope regarding scale and level of detail. The activity consists of a description of the historical sequence of the construction of the landscape considering meaningful events and situations which have determined its current condition. The information collected is then synthesized in a document containing its corresponding planimetric graphic references regarding the location of the area of study.

The methodological process is structured in three consecutive stages: identification, characterization and assessment. During the identification phase, the current landscape situation is discerned by noting and recognizing its structuring and defining components. Characterization entails perceiving homogenous areas in the region under scrutiny as to landscape character after a detailed study of every constituting component. Assessment is the result of the active and committed participation of the citizenry in appreciating and valuing the homogenous areas detected in the previous stage. Every phase is composed of specific goals, activities, guidelines, and graphic and conceptual products (file cards, maps, inventories, images, statistics, written documents, etc.). The three stages intend to define the condition, value, dynamics and tendency of the landscape through the analysis and

[3] The research team is made of Lucas Períes PhD. Arch. (director); María Cecilia Kesman, M.Arch. (co-director); Silvina De Lourdes Barraud, M.Arch.; Beatriz Ojeda, M.Arch. (researcher responsible). Collaborating researchers: Martha Bersano, plastic artist; Mauro Bianchi, industrial designer; Natalia Brizuela, architect; Cristian Castiblanco, architect; Cristina Chávez, geologist; Cecilia Eynard, biologist; Diana Perazzolo, agronomical engineer; Julio Rebaque, architect; Ricardo Riveros, landscape architect; Diego Alejandro Serra, biologist.

Fig. 3.1 Structure of methodological process

interpretation of the results obtained to define the landscape quality. These considerations recommend actions of preservation, recuperation and enhancement of the urban landscape (Fig. 3.1).

3.2.1 Identification

Identification is the first phase in the Townscape Catalogue construction. Landscape inspection is carried out observing the tangible factors of urban life as well as social, intangible factors (traditions, habits, lifestyles, and the like), broadening appraisal beyond the purely material.

We recognize two key references for the tasks of this phase: the manual method of direct observation of Litton (1972) and the project PAYS.DOC "Virtual Observatory of the Mediterranean landscape" (Junta de Andalucía, 2007), developed within the framework of the initiative INTERREG IIIB MEDOCC.[4]

At first, activities are oriented toward the definition of vantage points by determining observation spots to record panoramic views (through photographs) in summer and winter. An observation spot is a place where landscape is mainly regarded as to its accessibility and visibility conditions. Observation spots are areas that attract the largest number of onlookers and visual opportunities. Panoramic photography, in turn, is a specific type of wide-angle photograph in horizontal format corresponding to an ultra-wide POV.

A vantage point is a section of land visible from a given observation spot at a given moment in time. For the delimitation of a vantage point, the manual method of direct observation (Litton, 1972) is applied through a surrogate operation, based on the reading of panoramic photographs.

Observation spots are determined by carrying out field checks and revising location and aptitude via satellite images. In selecting observation spots, priority is given to sites where onlooker height and spatial width allow an ample view of the landscape and its compositional elements. The selection and number of observation spots must keep to a sequential logic with a connection between them. The distance

[4] European Regional Development Fund Programme for interregional cooperation in the European Union (2000–2006).

between observation spots should not exceed 600 yards (Litton, 1972) to enable their mutual visibility, i.e. the possibility of visual communication of both observation spots.

Photographs from every observation spot enable an analytic reading of the elements that make up every landscape by recording landscape components, i.e. those tangible and intangible elements that make up the landscape. The task consists of producing an analytic reading of the panoramic photographs to notice the landscape components that constitute every vantage point.

Our landscape component taxonomy distinguishes elements of different types: natural or anthropic, static or dynamic, and, finally, distant (those natural, cultural or mixed components visible from a distance, outside of the area of influence of a vantage point). Based on these five large groups of components, we propose a series of subcategories and types for their identification. The criteria that support this taxonomic scheme of landscape components are the result of an extensive study carried out with the participation of specialists in Biology and Agronomy to ensure the inclusion and distinction of the largest possible number of components from an interdisciplinary perspective. What follows is the taxonomic scheme developed (Tables 3.1, 3.2, and 3.3):

The panoramic photographs upon which landscape components are observed are also employed to produce other graphic products. The analysis of visibility plans entails interpreting visual schemes as layers that break up the observation of landscape at different depths. The analysis of environmental color encompasses regular hue schemes (in square pixel format) to obtain a color measurement percentage: Even though color is a dynamic landscape component depending on light and environment changes, it is necessary to apply a pixeling process to the panoramic photographs taken in similar environmental conditions.

This phase of the catalogue preparation results in a set of file cards containing the landscape components of every one of the vantage points, connecting written, graphic and photographic information produced from the data gathered in photographs and field observations. The support for the implementation of these Identification File Cards is taken from the PAYS.DOC project, which proposes a common system of recognition and observation, based on a large bank of images that allow the identification of different scenarios and situations.

File cards are made up of different elements: text describing vantage points, observation spots and cardinal direction; photographs of the location (observation spots) from the vantage point; the location map of the area of study; a vantage point map; summer and winter panoramic photographs; identification icons and landscape component number; color analysis of panoramic photographs; a visibility plan scheme; aerial photographs setting vantage point boundaries and reference numbers of recorded components, and the list of landscape component identification and singular or special landscape component photographs (Fig. 3.2).

The materials processed on the identification stage make up the initial platform in the construction of the catalogue. They document the natural and cultural as well as all tangible and intangible heritage that makes up the area of study. Although file

Table 3.1 Static landscape components

Static Landscape Components

GEOMORPHOLOGY	
Topographic relief mountainous, steep, terraced, flat, etc.	**Geographical feature** mountain, hill, cliff, island, valley, etc.
Hydrological relief river, brook, waterfall, pond, sea, dam, etc.	
BUILDING	
Individual building house, cabin, trailer, hut, etc.	**Management** company, office, call center, etc.
Collective building tower, condo, private neighborhood, hotel, etc.	**Transport** bus station, train station, airport, etc.
Business supermarket, shopping mall, store, etc.	**Culture** museum, theater, auditorium, library, amphitheater, etc.
Education university, school, kindergarten, etc.	**Entertainment** movie theater, café, restaurant, dance club, etc.
Religion temple, sanctuary, convent, hermitage, etc.	**Industry** industrial plant, factory, workshop, warehouse, etc.
Government state house, town hall, ministry, delegation, etc.	**Sports** stadium, sports club, health club, sports field, etc.
Healthcare hospital, clinic, dispensary, medical center, lab, etc.	**Green area** park, square, boulevard, botanical garden, reserve, zoo, etc.
Law enforcement police precinct, penitentiary, fire department, etc.	**Construction site** residence, business, education, religion, government, etc.
INFRASTRUCTURE	
Road street, highway, bridge, cycle lane, train track, etc.	**Energy** networks: power, natural gas, power station, etc.
Sanitation river, brook, waterfall, pond, sea, dam, etc.	**Telecommunications** television, telephone, internet (networks/antennas), etc .
FURNITURE	
Street lighting mast, column, lamp, reflector, etc.	**Recreation** slide, swing, seesaw, climbing frame, sandbox, etc.
Seating bench, stool, chair, etc.	**Artistic** statue, mural, bust, fountain, monolith, installation, mast, etc.
Containers trash can, dumpster, water fountain, mailbox, etc.	**Signage** billboard, advertising, screen, clock, graffiti, etc.
Traffic bollard, ford, traffic light, park meter, sign, etc.	**Fencing** fence, railing, wire fence, banister, hedge, green wall, etc.
Shelters bus stop, arbor, gazebo, etc.	
FLORA	
Tree over 50 ft. tall, up to 50 ft. tall, up to 30 ft. tall	**Succulent** cacti, fat
Shrub up to 15 ft. tall, up to 10 ft. tall, up to 5 ft. tall	**Vine** climbing, trailing
Palm tall, medium, short	**Herb** linear leaf, non-linear leaf

Source: Reproduced from Períes et al. (2016)

Table 3.2 Dynamic landscape components

Dynamic Landscape Components

Category	Component	Description
ACTIVITIES	**Management**	errand, procedure, formality, etc.
	Recreational	ride, stroll, reading-writing, drawing, gathering, etc.
	Business	buying, selling, etc.
	Religious	meditating, praying, etc.
	Sports	training, practice, etc.
	Transport	pedestrian, motor
	Education	formal, non-formal
	Tourist	guided visit, tour, free visit, etc.
	Maintenance	cleaning, repair, pruning, cutting, replacement, etc.
EVENTS	**Civic**	commemoration, celebration, parade, meeting, etc.
	Political	rally, march, protest, meeting, etc.
	Business	exhibit, show, fair, presentation, etc.
	Religious	celebration, funeral, procession, gathering, etc.
	Cultural	concert, show, exhibit, installation, etc.
	Environmental	reforestation, planting, festival, rally, etc.
	Sports	race, exhibition, game, tournament, etc.
VEHICLES	**Air**	glider, airplane, helicopter, etc.
	Water	barge, boat, cruiser, ferry, speedboat, etc.
	Rail	monorail, subway, tramway, train, etc.
	Land	car, bus, taxi, truck, semi, motorcycle, bicycle, etc.
FAUNA	**Vertebrate**	fish, amphibian, reptile, fowl, mammal
	Invertebrate	arthropod, mollusk, annelid
SIMPLE ORGANISMS	**Microscopic**	plankton, others
	Macroscopic	multi-cellular algae, multi-cellular fungi
MATTER	**Solid**	garbage, rubble, dry, vegetation, etc.
	Smell	vegetation, wood, urine, food, putrefaction, paint, etc.
	Liquid	oil, stagnant water, sewage water, fuel, etc.
	Sound	water, vegetation, fauna, people, siren, industry, music, etc.
	Gas	smoke, smog, gas, vapor, etc.

Source: Reproduced from Períes et al. (2016)

cards allow the execution of landscape characterization and assessment, they also constitute a detailed, in-depth study which may be used in other methodological frameworks and research projects pertaining the city, and even by other disciplines associated with its execution.

Table 3.3 Distant landscape components

Distant Landscape Components

Geomorphology	topographic relief, hydrological relief, geographical feature
Building	residence, business, education, religion, government, healthcare, law enforcement, management, transport, etc.
Infrastructure	rail, sanitation, power, telecommunications
Furniture	lighting, seats, containers, safety, traffic, shelter, recreation, artistic, signage, fencing
Flora	trees, shrubs, herbs, palms, succulent, vines
Activities	management, business, sports, maintenance, recreational, religious, tourist
Events	civic, business, cultural, political, religious, environmental air, rail, water, land
Vehicles	air, rail, water, land
Fauna	vertebrate, Invertebrate
Matter	solid, liquid, gas, smell, sound

Source: Reproduced from Períes et al. (2016)

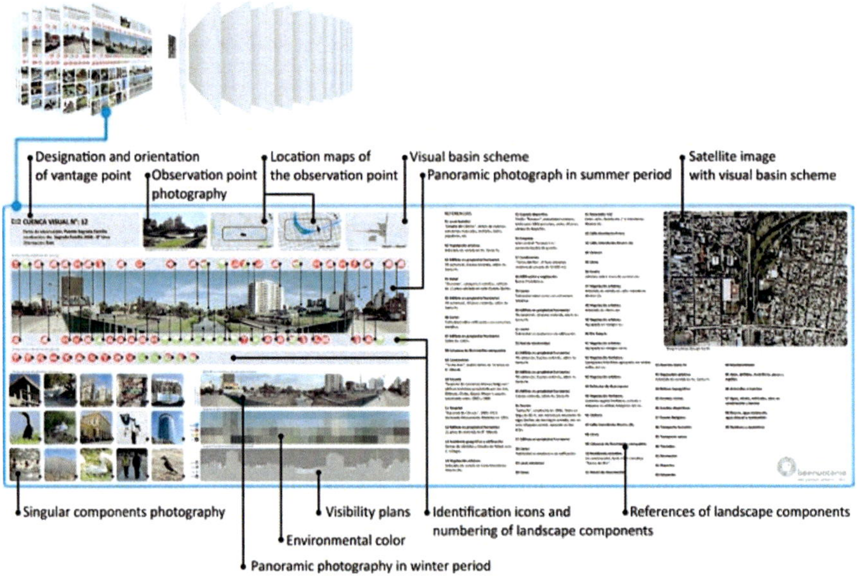

Fig. 3.2 Set of file cards of landscape components

3.2.2 Characterization

Characterization is the second phase in the Townscape Catalogue design. It is oriented toward recognizing homogenous zones within the area of study. Homogenous zones are outlined by their character, which is defined by the interrelation of their different landscape components.

The key reference for the tasks of this phase is the work of The Countryside Agency,[5] which proposes "Landscape Character Assessment provides a framework for describing an area in a systematic way. It lets different interest groups to make better judgments by knowing what's present and what is distinct, so any change can respect local character, or add to it, and even change it if what's desired" (Swanwick, 2002, 2). The Agency works on a map that divides England into 159 national character areas and allows a picture of the differences in landscape characteristics on a national scale. In our case, this strategy is adapted and developed for its implementation in the urban context.

After the identification work carried out on the first phase of the catalogue, what follows is the process of in-depth description of the distinguishing features of the landscape components in urban scale: This task is performed with the assistance of a system of charts that guide this characterization. Those charts are organized by the specific information of every component category. Firstly, the subcategory and type of the component is determined. Indicators are later disaggregated through a series of individual variables to deepen the description of every landscape component.

For example, on the latest research project, the second instance of townscape catalogue on Suquía river in the city of Córdoba, 6.650 components are analyzed at 27 vantage points and processed with an average of 15 indicators and an average of five variables each. The amount of data to process makes it necessary to use information technology to perform that task. For that reason, software has been specifically designed to that end.

Once data is fully uploaded, the software records the number of indicators and variables for every component category at a given vantage point. The subsequent processing of that information compares two adjacent vantage points. Such comparison renders the number of matching or different types – indicators and variables over the total number of components that make up the vantage point – resulting in rates of change. Thus, a comparative reading of different pairs of vantage points throughout the area of study indicates the recurrence or variability of the aspects that characterize the components by category (Fig. 3.3).

Rates of change in types and indicators are recorded on a cell chart integrating all 14 categories and the whole number of vantage points within the area of study. The rates of change are then gradated on a scale of 10 with 10% intervals. Every cell is given a color according to the gradient processed by the software to visually facilitate its reading.

[5] As part of Natural England, The Countryside Agency is a public body of the Departments of the United Kingdom Government, formed in October 2006.

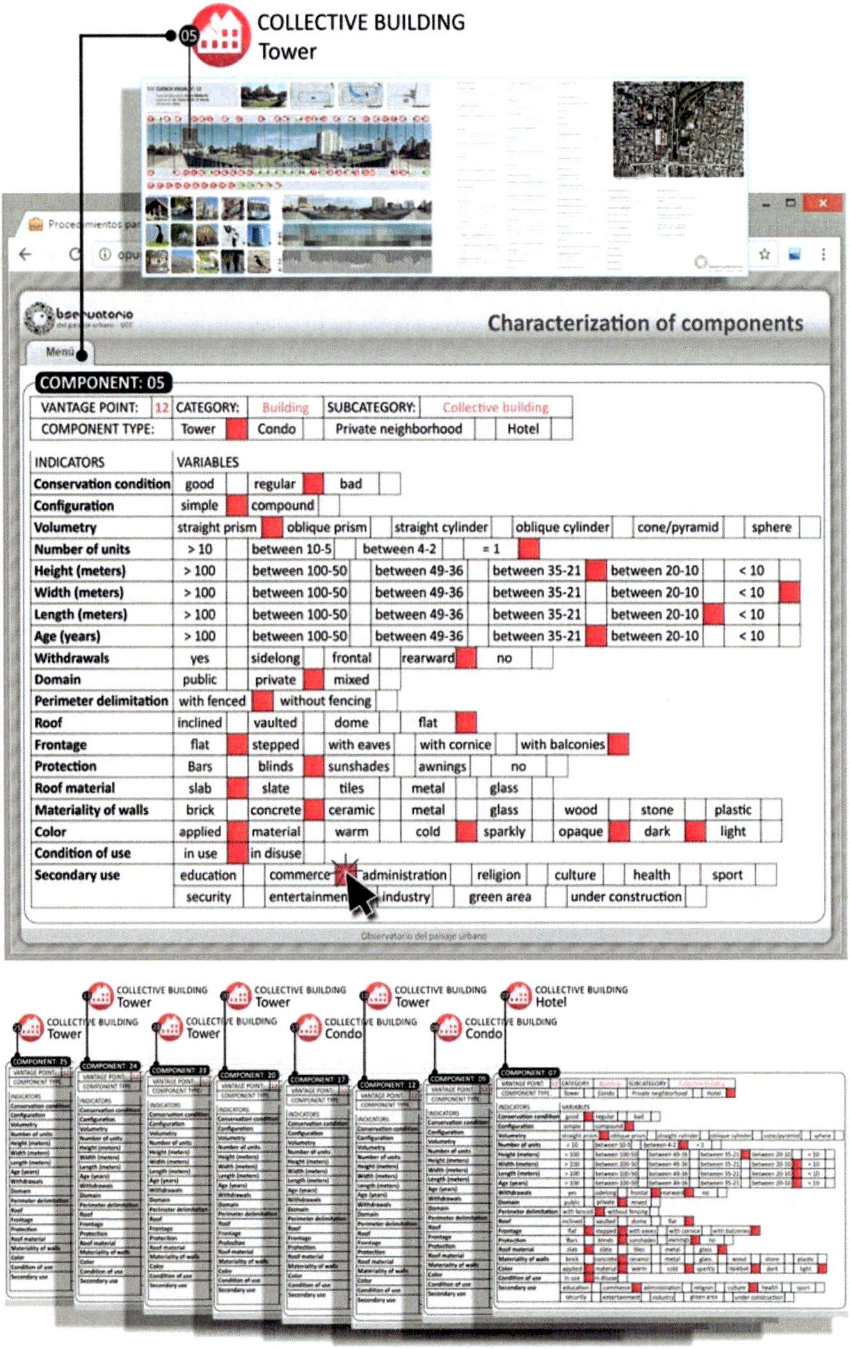

Fig. 3.3 Characterization tables of landscape's component category: collective building

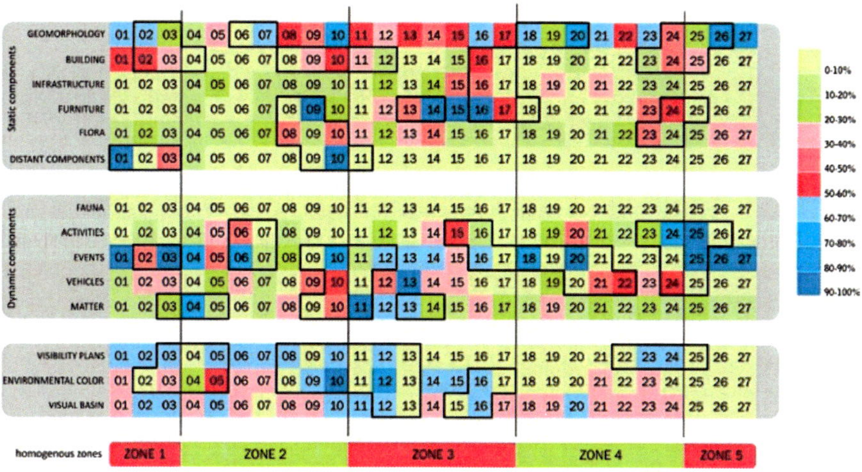

Fig. 3.4 Synthesis of homogeneity conditions

Combined with the environmental color studies, visibility plans and vantage point calculations, the resulting chart, made up of numbered cell columns (corresponding to every vantage point) and rows corresponding to categories of landscape components, shows the clusters of cells where situations or change are concentrated. These data sets are defined by identifying contiguous cells with a color gradient difference greater than or equal to 40%. In this way, it is possible to outline the axis that notes the largest amount of change between vantage points, defining the boundaries of areas which preserve homogenous conditions (Fig. 3.4).

Data processing following this system allows circumscribing the character of every vantage point, which is – through consistency or difference – associated in the definition, delimitation and characterization of homogenous zones within the area of study. A homogenous zone is composed of the specific combinations of various vantage points whose landscape components share similar features and confer a marked idiosyncrasy upon the rest of the region.

On this phase of the catalogue design, the in-depth study carried out results in the definition of areas with homogenous features which are transferred on a planimetric outline of the area of study. The character of every area is particularly developed on written documents including quantitative and qualitative information, with special emphasis on the most relevant categories which result from the interpretation of the synthesis scheme of homogeneity conditions.

3.2.3 Assessment

Assessment is the third phase in the Townscape Catalogue design. The main objective of this stage is evaluating landscape on the homogenous areas that result from considerations of landscape quality. Assessment consists of two instances: public consultation and assessment per se.

Fig. 3.5 Participants in the public consultation

Key references for the tasks of this phase are: The work of The Countryside Agency and the Landscape Observatory of Catalonia with different techniques and specific tools of social and community participation.

Firstly, the assessment and examination itself of homogenous areas on the part of landscape onlookers must take place. Active and committed citizenry participation constitutes a substantial contribution in assessing landscape. It is key to learn the impressions, observations and opinions of landscape specialists, interdisciplinary professionals and permanent or temporary residents to make an informed interpretation of the data collected in the previous phases (Fig. 3.5).

In the assessment of landscape, there are two survey techniques involved. The citizenry is surveyed on site by asking passers-by at every observation spot in homogenous areas (e.g. 450 people on the latest Suquía river research project). Interviewees respond on the positive and negative aspects of the landscape regarding what they are experiencing at that given moment as well as considering aspects related to the experience of living in that urban space in other circumstances and at other moments. In turn, there is an online survey directed at a representative group of landscape specialists (in different disciplines such as architecture, agronomy and biology; 19 in all in the current research project), and other multidisciplinary professionals (such as company managers, architects, artists, film-makers, journalists, accountants, graphic and industrial designers, historians, editors, engineers, medical doctors, psychologists, publicists and urban planners; 35 in all on the latest Suquía river research project). Using this technique, which is repeated on every homogenous area, interviewees respond on different issues pertaining the landscape. The main goal is to obtain quantitative and qualitative information through the observation of photographs and plans of every area, which initially serve a guiding purpose, but later trigger personal reflections, experiences and visions, from the standpoint of every interviewee's discipline in every case.

Secondly, the assessment of findings is achieved as a result of the three-phase process developed in the construction of the Townscape Catalogue. Findings are geared toward the understanding of the landscape form, dynamics, values and tendencies of every one of the homogenous areas, as well as the aspects, relations and indicators that explain that condition. This activity allows turning the assessment of the landscape into **Considerations of Landscape Quality** – proposing **preservation** actions on the townscape (elements, dynamics and relations which is necessary to protect, maintain or preserve), **recuperation** (components that demand sanitation or repair and pertinent regulations), and **enhancement** (elements, dynamics and aspects to improve or increase) – in the form of standards or guidelines directed at planning.

3.3 Possible Derivations of This Method Towards Urban Resilience

Urban resilience presupposes the capacity for a soon recuperation of urban systems in case of possible impacts which the city – as a complex system – may undergo. The strategy and methods proposed in the Townscape Catalogue may contribute to the construction of tools for surveying, studying and enhancing "potential areas of resilience," while it may also assist in recognizing townscape as a key holistic construction in the predefinition of areas which may be sensible to "urban-landscaping tolerance" tests.

The task of specification, typification and consolidation of every one of the three phases that make up the Townscape Catalogue design constitutes an antecedent for methodological replicability as to data management and articulation. Its detailed, in-depth system of identification and characterization allows multiple dimensions and levels of information analysis that the catalogue integrates, but could lend itself to other objectives and perspectives to guide its method, tools and procedure protocols partially or individually.

One of the features of the Townscape Catalogues lies on the definition of homogeneous landscape areas where it is possible to define identity codes for a specific part of the city. As an instrument to recognize the character of the landscape, the catalogue may entail a path to the construction of "flexibility and resilience patterns" of a part of the city, "buffer" sectors for the recuperation, mediation and endurance of systemic urban impacts, and the detection of landscape components which are apt or potentially prone to articulating the organic operation of flexible urban areas.

The identification phase and its graphic and conceptual by-products make up a platform which may enable the visualization of urban components (services) with the capacity of response, adaptation, recuperation and learning whether individually or as part of a network of "urban resilience objectives." In turn, it would also be possible to create a broader, specific taxonomy of the 'components of urban resilience."

The characterization phase offers concrete results: recognizing homogenous areas. A possible transference toward the construction of "indicators of urban recuperation capacity" may be a consequence of this phase from the resulting redefinition of criteria of data processing and reading. Completing distinguishing features by component; comparing characters to determine convergent or divergent conditions in vantage points, and generating pattern codes to arrange data reading are deemed as effective procedures in the definition of guidelines to establish boundaries in vulnerable areas, protected areas and opportunity areas regarding "urban impact and resilience."

The assessment phase integrates three interrelated products: public consultation exercises (to passers-by, professionals, landscape specialists), assessment of homogenous areas, and considerations on landscape quality. Apart from the concrete contribution of evaluation as a catalyst of the work done on the three stages of this strategy, every product may be oriented toward communication and interaction with the community, professionals and specialists in terms of "urban social resilience." Its main objective may be geared to informing about the results on "potential resilience areas" and "urban resilience objectives" as well as promoting identified communal urban practices which may contribute to raise awareness on resilience processes and available resource usage to develop projects of multidisciplinary adaptation, recuperation and participation.

3.4 Conclusions

In the process of successive research projects, it has been possible to produce two main contributions, one which is wide-encompassing and one which is specific: the consolidation of a method for the execution of Townsape Catalogues, and a contribution to further knowledge on the landscape of the city of Córdoba. In both cases, the knowledge created may contribute to new urban planning actions by state-run and private-run institutions, to further academic work, and to the development of future research. It is important to note that concrete measures of urban preservation, recuperation and enhancement demand an active position on the part of government institutions, a fact that exceeds both the objectives and reach of our research.

The purpose and execution of the methodological instrument proposed, the Townscape Catalogue, entails demonstrating the feedback processes achieved in transformative situations and city planning and organization actions. These are processes that foster reflection upon the work of different participants (governments, communities and universities) facing issues related to the preservation of landscape, environment, natural and cultural resources, biodiversity, and the living conditions and wellbeing of the citizenry.

Landscape sustainability in the city requires a system-oriented, multiple, open look upon the behavior of the different components that define and give structure to the city landscape, and which grant it a character and value which is recognized by the community. The proposed method responds to these demands from an encompassing perspective where landscape is deemed experientially, holistically and temporally involving the largest possible number of components – with their singular conditions and attributes – and it even includes the opinions of different participants in an instrument specifically designed. This instrument works on a section of the city observing the efficient performance of sustainable urban development policies and the prevention of risks and threats that affect urban landscape.

As an open research field, urban resilience may encounter conceptual and methodological foundations in the Townscape Catalogues from different aspects, outlooks and actions: Firstly, the transdisciplinary approach to face urban development and the integrated vision of the multiple and diverse subsystems that make up the city; secondly, the multiple procedural tools and protocols to carry out landscape studies and the multiple dimensions and levels in the analysis of the landscape components of the urban ecosystem, and finally, the integration and participation of multiple participants who provide their diverse perspective in the assessment of landscape.

References

Busquets, J., & Cortina, A. (2009). *Gestión del paisaje, manual de protección, gestión y ordenación del paisaje*. Ariel.

Council of Europe. (2000). *European landscape convention*. Council of Europe.

Cullen, G. (1961). *Townscape*. Reinhold.

Di Pace, M., & Caride Bartrons, H. (Eds.). (2004). *Ecología de la Ciudad*. Prometeo-UNGS.

Junta de Andalucía. (2007). *Observatorio virtual del paisaje mediterráneo*. Consejería de Obras Públicas y Transportes.

Litton, R. B. (1972). *Aesthetic dimensions of the landscape*. Johns Hopkins Universisty Press.

Lynch, K. (1960). *The image of the city*. MIT.

Naselli, C. A. (1992). *De ciudades, formas y paisajes*. ARQUNA.

Observatori del Paisatge de Cataluña. (2005). *Catálogos de Paisaje*. Observatori del paisatge. Retrieved October 10, 2015 at www.catpaisatge.net/esp/catalegs.php

Períes, L., Ojeda, B., Kesman, M. C., & Pedrazzani, M. J. (2011). *Nuevas Estrategias de preservación y potenciación del paisaje*. editorial EDUCC.

Períes, L., Ojeda, B., & Kesman, C. (2012). *Catálogo del paisaje del río Suquía en la ciudad de Córdoba* (Vol. 1: centro y peri-centro noroeste). EDUCC.

Períes, L., Ojeda, B., Kesman, C., & Barraud, S. (2013). *Procedimientos para un catálogo del paisaje urbano*. i+p.

Períes, L., Ojeda, B., Kesman, C., & Perazzolo, D. (2016). *Catálogo del paisaje del río Suquía en la ciudad de Córdoba* (Vol. 2: suburbano noroeste). EDUCC.

Swanwick, C. (2002). *Landscape character assessment: Guidance for England and Scotland*. The Countryside Agency & Scottish Natural Heritage.

Chapter 4
Building Resilient and Sustainable Cities Starting from the Urban Transport System

Emanuele Bellini, Paolo Nesi, Cristina Martelli, Evangelia Gaitanidou, Francesco Archetti, Antonio Candelieri, Jan-Paul Leuteritz, Pedro Ferreira, and Laura Coconea

Abstract The concept of resilience is emerging as one of the hot topics in several sectors. One of the most challenging sectors is that of the critical infrastructures which are present in our cities and involve the day life of everybody. Among these critical infrastructures probably the most prominent is the urban transport system since it is typically used as a resource for many other social and economic activities, and also to rescue and move people in the case of threat. In this context the European Commission has launched a number of projects with the aim of defining the European Resilience Management Guidelines, ERMG. Among those projects

E. Bellini (✉) · P. Nesi
DISIT Lab, Information Engineering Department, University of Florence, Florence, Italy
e-mail: Emanuele.bellini@unifi.it; Paolo.nesi@unifi.it

C. Martelli
Statistics, Informatics Application Department DISIA, University of Florence, Florence, Italy
e-mail: Cristina.martelli@unifi.it

E. Gaitanidou
Hellenic Institute of Transport/Centre for Research and Technology Hellas – HIT/CERTH, Thessaloniki, Greece
e-mail: lgait@certh.gr

F. Archetti · A. Candelieri
Consorzio Milano Ricerche, Milano, Italy
e-mail: archetti@milanoricerche.it; candelieri@milanoricerche.it

J.-P. Leuteritz
Fraunhofer IAO, Stuttgart, Germany
e-mail: jan-paul.leuteritz@iao.fraunhofer.de

P. Ferreira
Instituto Superior Técnico, Lisbon, Portugal
e-mail: Lisboapedro.ferreira@centec.tecnico.ulisboa.pt

L. Coconea
SWARCO Mizar, Turin, Italy
e-mail: laura.coconea@swarco.com

O. F. González Castillo et al. (eds.), *Urban Resilience: Methodologies, Tools and Evaluation*, Resilient Cities, https://doi.org/10.1007/978-3-031-07586-5_4

working on ERMG, the "Resilience management guidelines and operationalization applied to Urban Transport Environment (RESOLUTE)" project is focussed on the resilience aspects of urban transport system, and contributes to the global scenarios creating the guidelines and putting them into play (e.g., operationalize) with a number of real executable tools connected with the urban data by following a big data approach. This chapter is focused on presenting the state of the art of resilience conceptual structure for Urban Transport System (UTS) and the main aspects of RESOLUTE EC H2020 project.

Keywords Resilience engineering · Adaptive capacity · Urban transport system · Big data · Smart city

4.1 Introduction

Urban areas across the world are currently faced with the need for innovative management solutions that effectively address the increasingly complex nature of threats relating to both man-made and natural disasters. On the one hand, interdependencies between critical urban infrastructures and services are being rapidly intensified. This means that apparently local and small events may produce severe impacts across many other urban assets and even generate large-scale system failures. On the other hand, according to (IPCC, 2012), the climatic extremes may intensify or become more frequent and there are currently no scientific methods available to precisely predict the evolution and their spatial distribution.

Thus, increasing severity of weather related events renders climate change issues one of the most critical concerns for all city officials and urban service stakeholders, in particular those operating in critical infrastructures, such as energy and transport. In addition, the trend for growth of urban population seems to remain more or less steady, whilst resource scarcity also intensifies. This considerably increases the pressure to provide suitable preventive and mitigation actions against such large-scale failures.

The notion of resilience is rapidly emerging as a viable path towards successfully coping with the complexity of the challenges faced by urban areas. According to (Linkov et al., 2018), it has become prevalent among scientists, engineers, and policymakers in various socio-ecological fields (e.g., ecology, urban planning, flood protection, drought management) and across public domains (e.g., city managers, state, regional, and federal officials). Key stakeholders within industry, government, and society-at-large consider its application to problems such as disruption from climate change like Sendai Framework (United Nations, 2015) or the challenge of ecosystem management.

Resilience is commonly interpreted as different combinations of response, recovery and adaptation abilities, among others. Within a relatively short period of years the concept of resilience has known a wide range of different definitions and applications. Despite this diversity and the considerable amount of work dedicated to the subject, many of the shortfalls identified in more traditional approaches, namely relating to risk management, remain poorly addressed.

Within the scope of an EU funded project (RESOLUTE), (Bellini et al., 2016), an approach on urban resilience has been developed focusing on Urban Transport System. This approach starts by a new perspective on resilience and pursues the implementation of innovative solutions towards an effective application of UTS resilience management practices and tools to sustain urban resilience.

4.2 Urban Challenges in Global Changes

As urban population continues to grow across the world, the critical nature of the social and economic activities that cities support also tends to increase. According to the Word Bank[1] more than half of the global population lives in cities, and this number is expected to increase to two thirds of the global population by 2050. Demand for services in urban areas is therefore increasing exponentially, and the capacity of local governments to manage this demand is challenged. In fact, cities have always been at the core of knowledge creation, technology innovation and cultural change, among many other things. While this trend has marked the history of every city since their origins, in no other point in time has technology produced such profound impacts on the operation of all the critical services that cities provide and on urban life in general. Initially, the fast growth of information and communication technologies (ICT) has led to a widespread interconnection amongst people, which in turn produced strong interdependencies among every human social and economic activity. In recent years, as ICT becomes progressively embedded with automation solutions, the connectivity potential rapidly extends to what is currently defined as the "Internet of Things" (Höller et al., 2014). This fact has generated numerous interdependency related phenomena, the impacts of which are most profoundly felt in most major urban areas worldwide. While such strong and large-scale interdependencies enabled the pursuit of significant economic and social opportunities, they have also generated equally significant exposures to new threats, many of which remain poorly understood.

Overall, the complexity and fast pace changing nature of urban systems are generating many emergent challenges, in the face of which most currently used management and operational practices have demonstrated many shortfalls. The underspecified nature of operations in complex systems not only generates the potential for unforeseeable failures and cascading effects, but it also creates unexpected opportunities for intentional and unlawful acts of disturbance, as well as terrorism. "*Man-made disasters*" within recent years clearly illustrate the impacts

---

[1] http://www.worldbank.org/en/topic/urbandevelopment/brief/global-platform-for-sustainable-cities

of complexity and the underspecified nature of many operations, namely the London Ambulance Service failure (Finkelstein & Dowell, 1996), among others.

As discussed by Little (2010), the breaking of a water main can rapidly escalade from local impacts such as a street closure, to a major sinkhole that could be at the source of failures in other sub-systems such as natural gas or power supply. In return, such failures could result in fires, which would become difficult to fight due to accessibility difficulties and perhaps even to low water pressure caused by the initial leakage. However, simple this example might be, it clearly illustrates how systems often generate resource-based interdependencies that can emerge as complex operational and management challenges. An issue on UTS may have similar or worst impacts since it affects all the human activities (business, services, health, etc.). Challenges to urban areas are also rapidly emerging from the domain of climate change and the increasing potential for natural disasters that threatens many cities across the world. The increased severity and frequency of extreme weather events and the overall instability of climate seem to be an inescapable trend, in the face of which many city infrastructures and services have shown to be underprepared. Natural disasters tend to generate disruption on a much larger scale and often with more profound structural impacts. As multiple infrastructure and services are severely affected, response and recovery capacities also tend to be weakened and rapidly depleted. On the one hand, local resources may be either insufficient or their deployment impossible (for instance due to the lack of operational transport), and on the other hand, deploying resources from unaffected areas into affected ones is often made much more difficult. Additional care and work is currently devoted to territory management and urban planning requirements. However, modernising urban infrastructure in general and the UTS in particular, often requires substantial reconstruction interventions that high costs render unrealistic (e.g. tunnels, bridges). Thus, the adoption of more suitable designs and urban planning are often significantly constrained. Such issues become particularly complex when city historic and cultural heritage are at stake, as centenary architectural features pose additional challenges. For instance, has been proved that the vibrations generated by vehicles moving around the Florence Cathedral affected the stability of the Brunelleschi's dome. According to these evidences, new mobility solutions needs to be designed to reduce the risk while maintaining the level of mobility and place accessibility. Overall, whilst endeavouring to preserve their heritage, namely cultural and architectural traits, cities are confronted with the need to adapt to new trends of population activity and to increase severity of weather related events.

4.2.1 City Issues Affecting Resilience

In particular, according to (Bellini et al., 2016b), a city exhibits a number of challenges and issues because of its complexity. The following list emerged from the social dialogue exercise carried out during the 1st RESOLUTE workshop in

Florence, 2015, with the city actors (first responders, critical infrastructure managers, scientists, as well as business and citizens representatives):

- The presence of multiple-decision-makers fragments and sometimes prevents a clear-cut role definition of responsibilities for known and unknown risks.
- The supply chain interdependency issues are not properly taken into account by actors and stakeholders. Cooperation initiatives remain feeble and bounded by contractual, legal or otherwise more formal constrains.
- Citizens have their own conflicting micro-opportunistic behaviours, different risk perceptions, beliefs, skills, etc., that affect the city's resilience.
- Heterogeneous data sources (environmental sensors, traffic flows, social network, mobile apps flows, etc.) with different data delivery rate (ranging from real-time to static), quality, reliability and semantics.
- Slow in considering people as active actors of the urban socio-technical system and weak citizen engagement in decision-making.
- Difficulties of optimally managing the scarcity of resources in terms of first responders, goods and tools available before, during and after emergency and difficulties to have a list of priority actions.
- Lack of a coordinated multi-channel communication strategy and situation-aware communication delivery tools (e.g., localised and personalised early warnings, installation of variable messaging panels) resulting in potentially conflicting messages.
- Prevalence of management practices based on centralised control and prescriptive procedural approaches that do not take into account high variability and unpredictability of today's complex operations.
- Common attitude of the authorities to neglect the preparation and adaptation phases in favour of the absorbance and reaction phases with an over-specification of procedures and incremental development of rules that reduced the much needed flexibility (local adjustment).
- Weak population preparedness against unusual extreme events and wrong perception about their recurrence probability and potential effects

A survey conducted during the Mugnone Civil Protection exercise 2016 in Florence among the citizens living in the area of the exercise has revealed the following preliminary results: only 20% of responders stated that they feel prepared to cope with an emergency; and 34% of the citizens claimed that they tend to follow their own heuristics instead of the official communications during the emergency. Only 6% of the responders indicated their willingness to adapt their behaviour according to the instructions provided by the authorities. Another interesting result of the questionnaire is related to the utility of the civil protection exercise in the area, as perceived by the citizens. 86% of the responders did not consider such exercise useful to increase their preparedness and safety.

Such preliminary results reveal the existence of a critical issue at the social level towards the authorities devoted to deal with emergency situations. In turn, this status leads to apparent contradictions. In fact, citizens overestimate their own capabilities to cope with an emergency, leveraging personal experiences, while they admit the lack of preparedness. This result may suggest that citizens consider their low level of preparedness more reliable than the authority's capability to respond and manage the event.

It is worth to notice that during the interviews, several citizens complained that they were not warned about or involved in the exercise, despite that an extensive door-to-door campaign was undertaken by the civil protection personnel some days in advance. However, even if it is possible that the magnitude of the resulting score obtained in the question about the usefulness of the exercise can be boosted by a "reaction to a missed advice", the trend confirms the existence of a community and social issues that need to be addressed in order to enhance urban resilience.

4.3 Urban Sustained Adaptability Conceptualization

The diversity among definitions of resilience may lead to confusion around the implementation, operation and monitoring of resilience strategies (Larkin et al., 2015). The concept of resilience covers many different matters and is used across many different scientific domains. Within the field of resilience engineering this notion is defined as "*the intrinsic ability of a system to adjust its functioning prior to, during or following changes and disturbances, so that it can sustain required operations under both expected and unexpected conditions*" (Hollnagel, 2011). Because, it is based on the adjustment prior, during or after events, this concept must encompass a certain timescale, which underlines its dynamic nature. Thus, resilience is a process through which a balance between safety and efficiency is achieved and maintained, rather than a quality or condition of a given sociotechnical system. This balance must be built around as much efficiency as possible, maintaining operations close to the limits of system capacities and making the most of the resources available, whilst devoting enough attention and resources to safety as to avoid exceeding system capacities.

Urban resilience is thus the ability of a city, perceived as a complex sociotechnical system, to recognize when its performance variability is unanticipated and falls beyond its usual competence and adaptations (Woods, 2006) and to monitor and handle its resources and adaptive capacities needed to cope with such unwanted performance variation.

In line with the resilience engineering approach, the potential for resilience to emerge from system performance is assessed based on the "four resilience cornerstones" (Hollnagel, 2009):

- Knowing what to do (Respond)
- Knowing what to look for (Monitor)
- Knowing what to expect (Anticipate)
- Knowing what has happened (Learn)

Thus, resilience focuses on sustaining the capacity for a system to adapt in the presence of continuous changes. Generating, maintaining, and deploying adaptability processes relies upon the allocation of a wide range of resources and at many different system levels and time scales. As such, adaptability capacities are intrinsically related to the level of resources that a city can allocate and its ability to manage these resources in view of specific adaptive cycles. An adaptive cycle can be described based on the basis of the four stages of event management cycle that a system needs to maintain, in order to be resilient, see Fig. 4.1 (NAS, 2012):

- Plan/Prepare: Lay the foundation to keep services available and assets functioning during a disruptive event (malfunction or attack).
- Absorb: Maintain the most critical asset function and service availability while repelling or isolating the disruption or working in a degraded mode.
- Recover: Restore all asset function and service availability to their pre-event functionality as fast as possible.
- Adapt: Using knowledge from the event, alter protocol, configuration of the system, personnel training, or other aspects for further improvements.

Adaptive cycles cannot be dissociated from system performance variability, as they are simultaneously the mechanisms that systems use to cope with variability and an important source of variability themselves. In the same way that resources are inherently scarce, also the capacities for adaptability are limited. The challenges to system resilience reside then in the ability to understand and monitor resources and the capacities that they provide towards coping with both expected and unexpected amplitudes of performance variability.

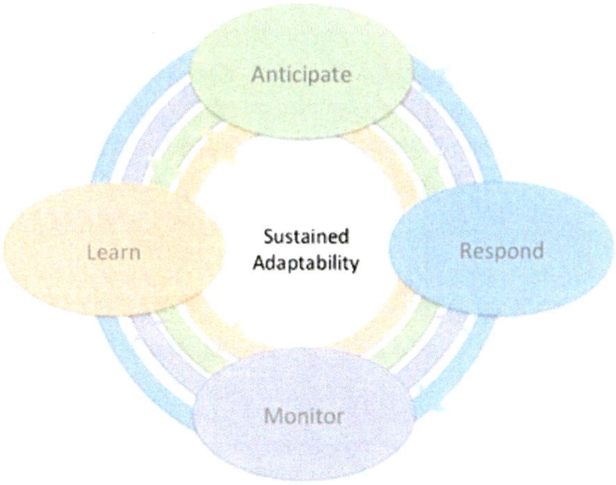

Fig. 4.1 Sustained adaptability cycle

4.3.1 City Assets

According to the RESOLUTE conceptual framework (Ferreira & Simões, 2016), from a resilience perspective, understanding system interdependencies and measuring adaptive capacities must take into account urban resources availability and their allocation (Bellini & Ferreira, 2018). Three fundamental sets of resources and assets must be considered in the urban context:

- **Human/social resources** include technical and non-technical skills of individuals, their heuristics, expertise and competencies, as well as cognitive resources, social characteristics (e.g. cohesion), cultural aspects, etc., all factors that may influence the decision-making processes. In fact, the human factor, characterised by behaviour based on limited (personal) knowledge of the status of the event, heuristic, physical skills, emotion, emulation, etc., is usually not adequately considered. The human behaviour during a disaster represents a critical and often unpredictable variable that can affect the operational side of the transport system (traffic jam, accidents, etc.) as well as the physical infrastructure side (too many vehicles and people on a bridge, etc.). In fact, actions based on wrong assumptions, panic, lack of knowledge of the ongoing scenario, lack of training, lack of coordination and right communication, and so forth can dramatically increase the amount of damages in terms of victims and exposures destroyed. It is clear that, the social dynamics and the human factor should be included in the resilience conceptualisation. The levels of commitment of urban communities towards their own well-being and in particular the factors that may promote such commitment, are considered a fundamental resource within the scope of RESOLUTE.
- **Technological resources** and assets (keeping in mind the focus on urban transport) comprises infrastructures and services such as water and energy supply, Urban Transport Systems, Intelligent Transport System, health care, business layer, energy layer, communication layer, etc. Special attention should be devoted to risk prevention and mitigation, and maintenance-related resources and assets, as these tend to be both critical system operational enablers and constrainers. The technological resources can be considered as the primary physical structures, technical facilities and systems which are socially, economically or operationally essential to the functioning of a society or community, both in routine circumstances and in the extreme circumstances of an emergency (UNISDR, 2009: 8–9).
- **Organisational** resources include hierarchical structures and formal procedures and regulations, as well as logistics elements. Information use and communication must also be taken into account, as they constitute the key resource for every decision-making process and for communicating with the population involved.

In line with the resilience engineering approach, system resources should be managed in such a way that they produce sustained adaptability capabilities at urban level. This leads to considering the following fundamental issues for enhancing urban resilience:

- What capabilities need to be enhanced?
- Where and when are such capabilities needed?
- Capabilities of sustained adaptability towards what?

To answer to such questions, the following section presents the evidence driven approach adopted within RESOLUTE, exploiting the possibility offered by the City of Florence to acquire and make available a huge variety of data, thanks to its Smart City developing program based on Internet of Everything (IoE).

4.4 A Big Knowledge-Information-Data Approach

There is a consensus around the importance of generating system capacities towards enhanced adaptability and on the fact that such capacities must be based on a more flexible management of available resources. The growing presence of technology at every system level pushed by the smart city trend, which is in turn supported by the Internet of Everything (IoE) propagation, renders Knowledge-Information-Data (KID) one of the most critical resources and therefore, one of the key factors to be considered when addressing urban resilience (Bellini & Nesi, 2018). Information is the basis of the large majority of urban interactions, whether these take place as Human-to-Machine (H2M), Human-to-Human (H2H), or Machine-to-Machine (M2M). Improved flexibility and adaptability exploiting information technologies constitute a fundamental step towards enhanced systems resilience, particularly by better supporting the ability to cope with increasingly higher degrees of variability and uncertainty, both under "normal" and degraded operational modes. Data has been transformed into information and, together with data mining and semantic processing, fuels the new generation of decision support systems.

The main advancement here is represented by the usage of real data generated by daily activities in the city (cumulating knowledge about critical events, and normal conditions) to assess resilience with respect to classical expert judgment and/or simulated data approaches. In fact, the capacity to cope with expected and unexpected changing conditions can be assessed and quantified by monitoring the resources/assets available in the city, how they are allocated and consumed, and so forth. It is possible to support decision makers in taking better decisions, as monitoring can now be performed in real time, and information and knowledge precision, comprehensiveness and granularity can be increased, while improving the timing, as Early Warning tool. In turn, the decisions can be phrased and communicated benefitting of the same enhancements: precision, comprehensiveness, granularity and timing. In particular, citizens can now be reached through a 4R communication strategy (right person, at the right time, in the right place with the right message). The effect of such feedback can immediately be detected through environmental sensors and channels, so that adjustments aiming to reduce unwanted variability (e.g., people escaping in a wrong direction) can be actuated in time. This loop, enabled by new technologies and KID approach, implements what is called the Evidence Driven Adaptive Cycle and needs to be continuously performed (Bellini et al., 2017). In fact, city managers should monitor resources availability and adapt their decisions not only during an emergency but also in daily activities, in order to prevent or mitigate the cascading effects of unwanted variability.

Therefore, the implementation of a Smart City and the corresponding exploitation of the Big KID generated sustain the urban adaptability and resilience since they enhance:

- the **monitoring** and control capability, by improving the granularity and breadth of knowledge and awareness about the system status and dynamics, continuously collecting Big Data from heterogeneous data sources/streams and sensors as the GPS positions of people, their concentration, typical trajectories in the city, behaviours and sentiment through smart devices and social networks (User Generated Data), Open Data, data from environmental and other kind of sensors (e.g., traffic flows, hydrometers, air pollution, underpasses water level, people flow, temperature), and real time reports such as weather forecast, social media, and so forth;
- the **responding** capability by providing detailed and timely information to authorities on one side, and to delivering personalised, real time, context aware, and ubiquitous advices to the community;
- the **learning** capability by applying advanced analysis on Big Data, performing data analytics and data mining to extract knowledge and learn from the events occurred, reaction performed, etc. (e.g., identifying metrics and indicators which may allow you to set up of predictive models for early warning and/or selecting suitable reactions by using statistic or machine learning approaches);
- the **anticipating** capability by continuously supporting the assessment of vulnerability and identifying when the system operates nearer to safety boundaries, predicting behaviours and event dynamics, supporting evidence based decisions at strategic, tactic and operation level, thus moving ahead with respect to the current practices based on pre-simulated emergency scenarios (Woltjer, 2006).

4.4.1 The Challenge of Measuring Resilience

The identification of the city's resources and performance characteristics allows the development of a set of requirements for indicators and monitoring tools as a fundamental support for management and decision-making (Häring et al., 2017).

In fact, decision-making in modern city and society needs to be guided even more by evidence. A quantitative approach to resilience measurement, in fact, triggers alerts, sustains scenarios analysis, supports in finding recovery paths, and so forth (Bellini et al., 2016c).

However, resilience is a complex and multifaceted concept and a lack of a shared understanding about its nature (process, property, outcome, performance, etc.) prevents the possibility to develop indicators that can be understood and accepted by all the city's stakeholders.

Implementing a measurement system for city resilience assessment requires a clear understanding of the level of details and precisions intended to be achieved since it will impact on the resources (skill, costs, time) needed for modelling and data processing (Linkov et al., 2018; Bellini et al., 2019). Moreover, it has implications at political level because of its connection with the citizens' quality of life. Moreover, without a common and clear understanding of what an indicator actually means, the score obtained is prone to wanted or unwanted misunderstandings. Transparency is imperative in public administration, in this perspective, any kind of "semantic and linguistic gap has to be avoided. Informative noise, in fact, may seriously affect the overall quality of measurement results and hampers the figures exploitation in a political and social debate. Indicators require an agreement about their acceptance among all the actors involved, to achieve a clear and common understanding of the city adaptive capacity is and how can be monitored and quantified. In this respect a semantic reconciliation of the most ambiguous concepts and the identification of an official and widely accepted definition for them is crucial. The consensus driven development of community based semantic glossaries is a crucial step to bridge the unstructured language of politics, persons and institutions to pertinent indicators (Bellini et al., 2017, 2019).

For relevant indicators a complete documentation metadata information system has been developed (algorithm, how to read figures, bibliography, pertinent data sources) to sustain a deep and reciprocal understanding among the network nodes.

Such approach should result in a semantic-aware Statistical Information System, to govern indicator development according to the transparency imperative (Martelli & Bellini, 2012; Bellini et al. 2016a, 2019).

4.5 Operationalize the City Resilience Concept: The Urban Transport System Case Study

Enhancing resilience in the city requires the enhancement of one of the most important critical infrastructure: the Urban Transport Systems (UTS). This is considered mandatory for two main reasons: first of all, such systems provide critical support to every socio-economic activity and are currently themselves one of the most important economic sectors in Europe (EU, 2014) and, secondly, the paths that convey people, goods and information, are the same through which risks are propagated. Transport systems have thus developed a prominent safety and business critical nature, in view of which current management practices have shown evidence of important limitations. In essence, the majority of management practices are based on the assumption that systems and their operations can be fully understood and described. This does not take into account the underspecified nature of complex sociotechnical systems and the need to cope with fast pace changing environments.

UTS today have to cope with significant vulnerabilities: aging infrastructure components, continuously increasing concentration of populations at urban areas, increasing interdependencies among physical and cyber infrastructures, co-location of many transportation systems with large-scale and potentially hazardous production facilities, along with ever growing threats of climate change and terrorism. All these have created significant challenges for the critical infrastructure systems.

A framework for enhancing critical transportation infrastructure resilience could potentially serve as a roadmap for addressing some of these pressing global challenges. The concept of resilience, however, has been broadly used to characterize a system that recovers rapidly from a disruption in order to resume normal operation. And **resilience does not only involve recovery: UTS r**esilience is an overall concept, defining a complex transportation system that is able to **better withstand disruptions**. The transportation system includes physical, technical, social, and institutional elements that are all critical to resilience.

A resilience framework should be addressed as the framework within which it would become possible (by making the most out of existing capabilities and resources) to create or convert a transportation system into being more effective and much less vulnerable to disruption. A resilient transportation system has own robustness to withstand severe drawbacks; it is adaptable in order to respond appropriately to threats and can mitigate the consequences of threats through response and recovery operations. These three attributes—**robustness, adaptiveness, and consequence mitigation**—form the foundations of a resilient transportation system (Volpe Center, 2013).

RESOLUTE is based on the vision of achieving higher sustainability of operations in European UTS. The project recognises foremost the ongoing profound transformation of urban environments in view of ecological, human and overall safety and security needs, as well as the growing importance of mobility within every human activity. Within this context, RESOLUTE considers resilience as a useful management paradigm, within which adaptability capacities are considered paramount. Rather than targeting continuous economic and financial growth of businesses and market shares, organisations should develop the ability to continuously adjust to ever-changing operational environments.

4.5.1 Why Is the UTS Critical?

Within the urban context, transport systems are today challenged to respond to a wide range of mobility needs, whilst coping with severe constraints of many different kinds, namely geographical, environmental, safety and security-related. The UTS is a relevant case of an interconnected system where critical infrastructure and multi decision makers (people, civil protection, public administration, etc.) are involved. Moreover, the UTS is linked *to The Four* critical infrastructures (as defined within the European Programme for Critical Infrastructure Protection – EPCIP)

(EU, 2012) and following the approach suggested within the EPCIP, the concepts are based on a multidisciplinary holistic view of resilience for critical infrastructure.

The numerous links and interdependencies of UTS increase its criticality in terms of critical infrastructure resilience. In the case of an emergency, regardless of the system that it may address, the transport system is always affected, either by the emergency itself, or by the actions and measures to prevent or confront it.

Thus, when considering the above mentioned interdependencies, along with the complexity of the system as such and its high exposure to a variety of threats (from system malfunctions to climate disasters and terrorist attacks), managing resilience in the UTS becomes a multi-parametric task of primal criticality.

4.5.2 UTS Complexity

As indicated above, UTS is a system of high complexity. There are various factors that define and, thus, underline its criticality:

- *Large scale networks*: Especially in big cities or metropolitan areas, UTS is composed by networks of great volume and different nature (road, rail, bus lines, etc.). Managing each type of network greatly varies and the bigger the scale the most complex and difficult it is to effectively assure its proper operation. Moreover, impacts migration issues between network components, in case of a disruption in one of them, make this task even more challenging.
- *Multiple modes*: UTS potentially comprises most of the available transportation modes, including road (private vs public transport, cars vs two-wheelers), rail (surface vs underground), pedestrians and waterways. This characteristic alone is significantly raising the system's complexity, as the needs of each mode (and of each of its components) vary, both in normal operation and during an emergency.
- *Multiple operators – scattered responsibility*: This is directly linked to the previous point, as the existence of multiple operators in the system can be an additional factor of complexity in terms of responsibilities' distribution. This becomes more crucial during an emergency, as the management of system actors would require coordination in multiple and different levels.
- *Multiple industries*: A direct consequence of all the above is the involvement of a great range of related industries. Vehicle and building are the most evident ones, together with electronics, cyber technologies, fuels, commerce, etc. The list is long and the influence and impact of UTS to each of them is significant and multi-parametrical.
- *Public-private mix*: The fact that UTS is composed by both public (public transport vehicles, infrastructure, etc.) and private components (own vehicles, pedestrian movement) is another factor of its complexity, as there are different rules and means of controlling their individual operation, in addition to the overall system management.

- *Multiple recipients* (people, freight): The main recipients of the services offered by UTS are people and goods. In both cases the demanded service can be highly differentiated and unpredictable, in terms of its nature, frequency, duration, location, etc. Moreover, a disruption in the system operation could result in major social and economic consequences.
- *Critical to economy*: The criticality of UTS in the local, national and international economy is obvious, not only from all the above mentioned factors, but also due to the fact that the urban environment is the core of the economy and UTS is the primary means for any kind of economic activity to flourish and succeed to its targets, e.g. movement of goods, transportation of people to work, leisure, shopping centre, etc.

Thus, it is evident that dealing with UTS means dealing with a highly complex, multi-actor and multi-parametric system, whose management – especially in terms of resilience – is a critical and challenging task. This complexity is definitely a drawback for achieving absolute control of the system and, in fact, this is not the aim of resilience management. However, it requires the establishment of concrete structures and strong synergies, in terms of envisaging a sustainable and resilient operation.

4.5.3 Threats Exposure

Apart from its high complexity, UTS is a critical system in terms of resilience, also due to the fact that it is widely exposed to threats. Due to its multi-operational, multi-actor, distributed and multi-component nature, UTS's exposure to threats is really unlimited and characterized by high uncertainty and unpredictability. By the term "threats" it is considered any external event that may lead to the disruption of the normal system operation. For the case of UTS, such events are mainly deriving from man-made threats (e.g., human error resulting in accidents or operational faults, terrorist actions), major nature destructions (earthquakes, floods, climate change effects, etc.) or system malfunctions. Each of these sources of threats may impose a vast variety of events that the UTS would need to face; thus, enhancement of UTS resilience is aiming to secure the optimum level of system operation in a great variety of possible situations.

A very interesting and recent review on resilience and vulnerability analysis is proposed in (Mattsson & Jenelius, 2015), clearly reporting that there is a substantial literature on vulnerability analysis approach for transport systems, while the literature on resilience is less extensive. A consideration further supported in Faturechi and Miller-Hooks (2014) and Khademi et al. (2015) offer a comprehensive overview of transport system performance during disasters.

To assess performance of any infrastructure, the physical damage state of its components, usually predicted through fragility curves, is less important than the quantification of functionality losses, which are essential both for evaluating resilience and defining restoration strategies. In (Gehl et al., 2016) a component-based approach for the derivation of fragility functions and the effect of the multi-hazard fragility of bridges on the performance of road networks is investigated. This approach is important as it enables (i) the evaluation of the damage in terms of functional losses and downtime duration and (ii) the harmonization of the fragility models between several hazard types, while accounting for potential accumulation of damages. Although focus of the study on bridges, the proposed methodological framework is valid for analyzing other network elements such as tunnels or road segments and it is addressed in RESOLUTE's application framework.

Recent interesting studies and solutions can be mentioned such as Dehghani et al. (2014), that compare topological-based and system-based measures; Reggiani (2013) and Reggiani et al. (2015), who focused on how resilience and vulnerability can be framed, interpreted and measured, and their relationship with connectivity of the associated graph; Khademi et al. (2015) who provide a comprehensive approach to analyze vulnerability/resilience of a UTS with respect to a catastrophic event, in particular by analyzing changes in travel demand and behavior in the response phase of a disaster when emergency trips have to be prioritized and many roads are impassable.

4.5.4 UTS Resilience

Resilience design and management for UTS is nowadays a necessity, in order to effectively confront the difficulties arising from UTS system complexity and exposure to threats. The general principles that should be followed could be summarized in the following (US DOT, 2003):

- Prevent incidents within control and responsibility, effectively protect critical assets.
- Respond to events that cannot be prevented, mitigate loss and protect employees, passengers and emergency respondents.
- Support response to events that impact local communities, integrating equipment and capabilities seamlessly into the total effort.
- Recover from major events, taking full advantage of available resources and programs.

4.6 Work in RESOLUTE

To this aim, RESOLUTE proposes a set of European Resilience Management Guidelines - ERMG (Gaitanidou et al., 2016) addressing any Critical Infrastructure, which are "translated" in the context of UTS, based also on the theoretical background described in the previous sections. The functions that compose the operation

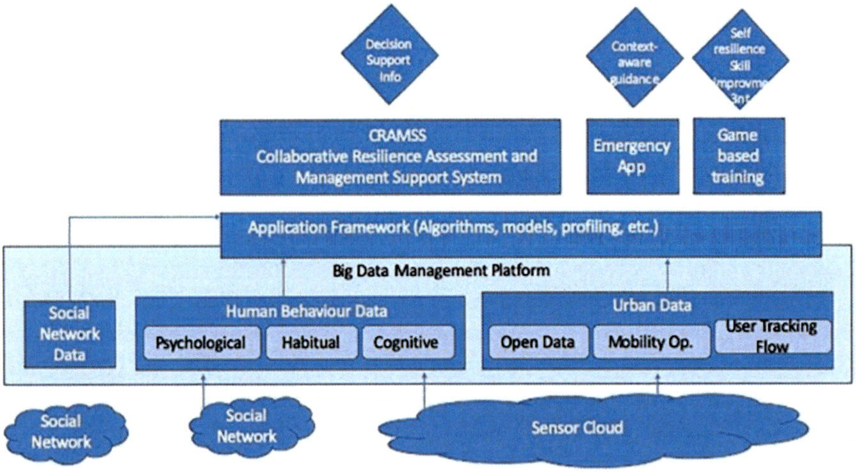

Fig. 4.2 Operationalization of RESOLUTE concepts

of UTS have been defined and the relevant guideline for each of these functions is addressed, resulting in more than 250 guidelines (18 functions × 13 guidelines categories) (Gaitanidou & Tsami, 2016). These guidelines are going to be operationalized with the use of a high level decision support tool, named CRAMSS (Collaborative Resilience Assessment and Management Support System) collaborating with a number of lower level decision support tools (namely, ITS, UTS, Fire Department decision support, energy control room, etc.), along with a Mobile emergency app and a Game base training app. All of them developed within RESOLUTE. The operationalization is taking take place in the two pilot sites and urban transport systems, namely the City of Florence and Athens Metro. The operationalization and testing with the use of the tools mentioned above feed back to the guidelines construction contributing to the final version of ERMG and ERMG for UTS, taking into account their application in real life critical environments, thus already performing a critical assessment of the guidelines, aiming to avoid possible drawbacks and gaps (see Fig. 4.2).

4.6.1 *Applying the KID Approach from a UTS Operator's Perspective*

Following the resilience approach requires empowering individual and locally distributed human actors to optimally apply their available resources to reduce critical variability in the functions of the UTS for example by creating buffer capacities. Thus, the UTS relies on human actors to deal with *dynamics*, *complexity* and *uncertainty* (Norros, 2004) that cannot be controlled on the basis of fixed rules and procedures. For this purpose, the user requires tools and an organizational context that allows doing so efficiently.

Dynamics refer to the fast changing status of single functions or aspects connected to the functions, such as outputs. The CRAMSS supports dealing with dynamics by providing constantly updated information in the shape of personalized widgets that join information from a variety of sensors and sources.

Complexity refers to the number of elements and interconnections that define a system, and thus the difficulty of predicting system behaviour. Dealing with complexity requires a top level overview of the system and its status, such as a representation of the system's functions and their interconnections. The CRAMSS features a representation of the status of the UTS, modelled as a FRAM, to decrease the effect of complexity and help the user understand and predict the behaviour of the system as a whole.

Finally, users have to deal with task *uncertainty*: the existence of unclear connections or links between what has to be done and the results that have to be achieved (Navarro et al., 2008). Following empirical findings, one should address three types of task uncertainty: unclarity of goals, new situations, and non-routine.

1. "Unclarity of goals": this type of task uncertainty is characterized by a lack of definition of objectives; achieving sufficient coordination among locally dispersed operators requires each of them to understand which objectives have been prioritised by the top level authorities. The CRAMSS allows for the top level decision maker to declare such priorities and consequently to allow each actor to see the broader picture.
2. "New situations" are characterized by unpredictable and fast-changing demands from the outside that the worker or team needs to adapt to. The CRAMSS responds to this demand by providing top level objectives and informing about the status of relevant variables, such as weather, traffic situation, the output of decision support systems, and social media analysis.
3. Non-routine refers to the methods and tools used by the operator to deal with tasks. The resilience approach responds to this type of uncertainty by demanding the same tools and methods to be used in normal conditions and emergency situations. This is realized in the CRAMSS: the functionalities available do not change depending on any emergency having been detected or not.

4.6.2 Big Data Management and Policy Issues

There are three types of data being collected and managed by the Big Data Management Platform and used by the CRAMSS: urban data, human behaviour data and social network data. There are four types of data being collected and managed by the City Big Data & Service Aggregator and used by the Data Analytics Semantic Computing layer up to the front-end applications (see Fig. 4.2). The City

Big Data & Service Aggregator (Bellini et al., 2014), collect and semantically aggregate knowledge and data coming from different sources among which:

Urban data include municipality open data, such as: structure of the city, seismic risk maps, hydrological risk maps, services, statistics, time series of major disasters, descriptors of structures such as schools, hospitals, streets, river level, weather conditions, position of Wi-Fi AP, locations of people aggregation facilities (such as: gym, schools, mall, social house, theatres, stadium, hospital). Some of these issues generate real time data such as emergency triage status of hospital, environmental sensors, etc. (Bellini et al., 2016).

UTS Big Data such as: description of the public transportation, busses poles and their timelines, taxi, parking areas and availability, metro status and position, cycle paths, restricted traffic zone, street direction and capabilities, traffic flow information, origin destination matrices for cars, traffic flow movements.

Human **behaviour** data may be either individual or group-based and include activity related and behavioural personal or collective profiles addressing **psychological**, **habitual** and **cognitive** aspects. These profiles may be extracted based on different kinds of sensors: Wi-Fi network, Bluetooth servers, traffic flow sensors as spires, TV-cameras, mobile cells from telecom operators, mobile Apps, etc., by using data mining, data analytics techniques, processing huge amount of data related to the single movements in the city. In addition, information through specialised surveys in order to assess negative human feelings during a post-crisis period. All these multidisciplinary and multimodal raw data need to be integrated in a common comprehensive format towards discovering meaning-bearing annotations (Bellini et al., 2014).

Another important input is data from **social networks**. A social network crawler can be exploited to manage and analyse all real-time data streaming from the citizens and the city infrastructure (Grasso et al., 2016). The crawler should be language independent utilizing multilingual thesaurus. Text processing and knowledge mining techniques should be used to discover hidden information. In addition to the dynamic data, an interoperable knowledge base contains cross sectors data that can be used to provide services to help the environment to become more efficient in disaster situations. Furthermore, the activities of data analytics and semantic reasoning are used to generate new knowledge that can be integrated into the interoperable knowledge base where cross sectors data are used to help improve resilience in situations of danger (e.g., data ingestion, mining and algorithms, computing models and recommendations) (Fig. 4.3).

These heterogeneous datasets have to be accommodated in a scalable and interoperable Knowledge Base (Bellini et al., 2014), which contains cross-sectors data that can be used to provide services to help the environment become more efficient in the event of a disaster. Furthermore, the Data Analytics Semantic Computing layer computes several elaboration to generate new knowledge (such as: extraction of typical human trajectories in the city, computation of the origin destination matrices at different time slots and week day, compute predictions about eventual city dysfunctions, compute sentiment analysis with respect to major city services) that can be integrated into the Knowledge Base, where cross-sectors data

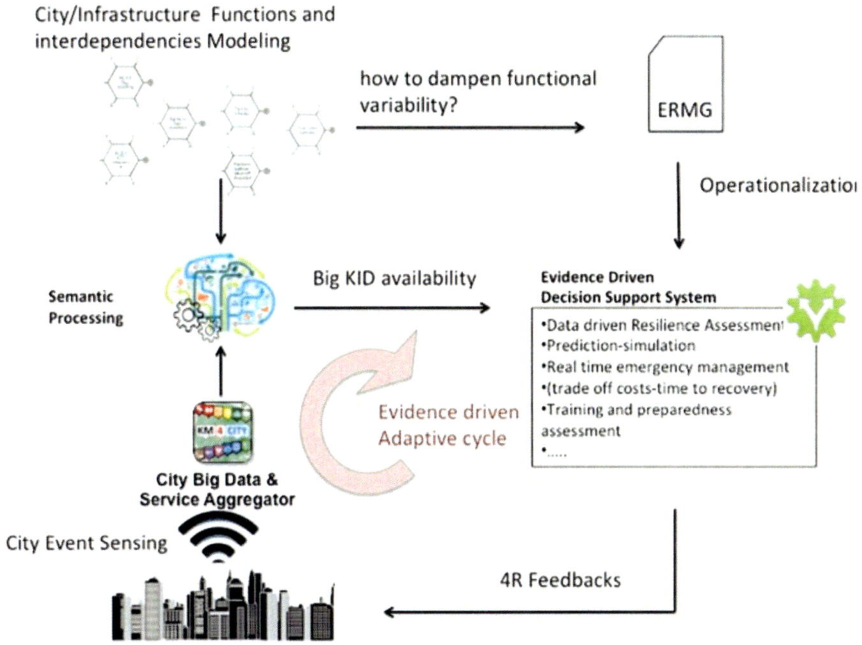

Fig. 4.3 Evidence driven adaptive cycle in urban context

are used to help improving resilience in situations of danger (e.g., data ingestion, mining and algorithms, computing models and recommendations).

RESOLUTE uses public open data, private data from the companies, and data coming from citizens requires a specific data management policy. The main aim of data processing in RESOLUTE is related to the assessment of risk and resilience. To this end, the resulting algorithms and data are delivered as confidential and restricted. In addition, a specificity of the RESOLUTE project is the exploration of social networks and technologies (e.g., Twitter) and open web forums, from which data will be collected and analysed to improve outbreak detection. Only the project partners have the access to such data, and global disease emergence related data rather than individual information will be exploited in the frame of the project. In any case, the partners involved in the corresponding tasks should make sure their activities comply with European Community legislation on personal data protection implemented in national law and regulations, i.e.: Directive 2006/24/EC of the European Parliament and of the Council of 15 March 2006 on the retention of data generated or processed in connection with the provision of publicly available electronic communications services or of public communications networks and amending Directive 2002/58/EC. RESOLUTE put in place also all needed actions to guarantee the security of data in relationship to the identified business scenarios with the following actions (and others that may come from changes in the context, national laws in the countries in which the demonstrators and validations actions will be performed):

- Appropriate licensing schemes for Data via the Data Licensing Tool in agreement with standards and national laws
- Conformity with data protection law
- Enabling access to the sensitive data only by qualified, authorized and authenticated users, and all the accesses and actions will be tracked and monitored.
- Enabling the configuration of RESOLUTE Challenges only by qualified, authorized and authenticated users, and all the accesses and actions will be tracked and monitored.
- Saving user profile data in secure locations and storages that can be accessed only by qualified, authorized and authenticated users, and all the accesses and actions will be tracked and monitored.

The RESOLUTE consortium will comply with European Community legislation on data export and protection to prevent the usage of the data for military and terrorism and in any case to trace the access to these data from any user and any title, tracking the access singularly and assigning personal authorizations, i.e.: REGULATION (EC) No 428/2009 of 5 May 2009, Official Journal of the European Union, setting up a Community regime for the control of exports, transfer, brokering and transit of dual-use items. For instance, privacy preserving rules are applied during the acquisition of all these datasets, allowing the extraction of cross-sectors correlations and the identification of risk mitigation strategies, especially in critical situations.

4.6.3 Application Framework

The Application Framework implements the models, algorithms and rules to mining and analysis available data mainly focusing on: managing user profiles and behaviors, analyzing cascading behavior in UTS modeled network, predicting in time the evolution of the network itself and estimate the resilience metric with respect to the degradation of service (effect) and not only node failure (both cause and effect). The model learn and "proactive" anticipate events that could lead to a failure of the UTS network and communicate to the monitoring and response components, hence assuring that the overall system can provide adequate response following predefined guidelines.

Results from graph theory and network science offer the possibility to develop and use tools for supporting decisions in UTS resilience management, including *preparedness*, *response*, *recovery* and *adaptation*. Although most of the research addresses vulnerability – usually related to decision-making during the pre-hazard phase – the application framework developed in RESOLUTE aims at supporting the operationalization of resilience concepts by offering network science algorithms and models for vulnerability analysis and more recent resilience evaluation.

With respect to vulnerability analysis, the RESOLUTE's application framework provides tools from the two distinct research lines – which had limited interactions: (a) topological and (b) system-based vulnerability analysis tools.

As for the topological approaches, the system-based ones model the UTS as a graph but also integrate further data in their analysis, such as demand and supply, resources and costs, as well as results from modeling and software simulation. Another important data sources that can be used by transport authorities in vulnerability/resilience analysis are related to "opportunistic sensors" like taxi GPS systems that are becoming increasingly available (Jenelius & Koutsopoulos, 2013). More important, system-based vulnerability approaches allow for a better definition of impact measures with respect to consequences of disruptive events. These measures may range from very simple ones (e.g. increase in travel time, increase in travel cost, cancelled travel options) over more general measures of accessibility to comprehensive economic measures of consumer surplus or financial impacts.

4.6.4 Collaborative Resilience Assessment and Management Support System (CRAMSS)

The Collaborative Resilience Assessment and Management Support System (CRAMSS) of RESOLUTE is a critical aspect of the resilient assessment and management concepts and methods included in the European Resilience Management Guide (ERMG) for the Urban Transport domain. CRAMSS adopts a highly joint approach taking into account intra and inter system interactions towards defining a resilience model for the next-generation of collaborative emergency services and decision-making processes. The **CRAMSS** is a Multiple Input – Multiple Output (MIMO) system having the capability of getting and analysing in real-time multivariate, asynchronous and/or pre-computed data.

Based on all the received input, the CRAMSS comprises an adaptive and constantly learning decision support system which can be capable to extract a set of possible resilience strategies in several layers of abstraction and subsequently it will select and implement the best ones according to situation. The output information of CRAMSS is communicated through situated displays (relevant stakeholders and the wider public), external stimuli (e.g., traffic lights) and/or users' devices providing visual/hearing and cognitive aids. The latter case also includes the emergency support **smart mobile app**.

4.6.5 Mobile Emergency App

The success of mitigation practices requires the collaboration of specialized personnel and citizens. It is important that the whole community 'is aware of the risks and worry to take action to prevent them'. Reactions with respect to incidents represent one of the greatest challenges in maintenance and emergency management. In most cases, the accessible information on the nature of the incidents is inaccurate as the needs to solve them; thus the personnel is inefficiently coordinated, informed neither on real conditions, nor on available resources. The logistics aspects related to the intervention and to the movement of personnel and patients are very relevant. Involved personnel need to have access at updated information and knowledge in the emergency and maintenance conditions. Therefore, mobile devices are mandatory tools for information access and to help sometimes in taking decisions. On such grounds, the Mobile Emergency App has to guarantee the access to any right and updated information in the needed time. The Mobile Emergency App aims at being a solution to guide personnel during maintenance and/or emergency conditions, that can help to reduce the time needed to react and to cope with organization and maintenance support, while facilitating communication, and indoor/outdoor navigation. The App is based on the formalization of protocol, the modelling of knowledge for navigation, the algorithms and a server device for integrated indoor/outdoor navigation. In this context, the application will be also connected to the CRAMSS framework and will support different end users roles (from emergency teams to individual travellers in accordance with the user profile used in each mobile device), expanding the concepts already tested in local emergency cases such as in Mobile Emergency in hospitals (Bellini et al., 2014b). The application is going to utilize optimally contextual information stemming from the smart device (e.g. location, environmental, etc.), which coupled with its role and other individual characteristics, and provide micro-tasks and effective inter-communication and data sharing (e.g., among rescuers).

4.6.6 Game Based Training App

Also the educational challenge has been faced in a non-conventional way, as it is impossible to answer to urban resilience global questions with partial answers, generated by specific disciplines languages and sectors: we need instruments able to speak to everybody, and to prepare to complex critical situations. The serious games approach has been adopted. The game approach, in fact transforms the critical elements derived by the functional analysis in decision points in which the player tests his/her knowledge and mastery of critical situations. A good serious game scenario allows to draw complex frameworks and asks for holistic solutions.

Game-based learning has become an optimal training tool for soft skills development since it fulfills the following five criteria:

- Compelling content
- Clear emphasis on practical application
- Interactivity and experimentation
- Genuine skills development through practice and feedback
- Motivation for people to learn and, above all, to complete the course they begin.
- Latent dimension recognition (like, for instance, to propensity to risk)

In the adopted approach a deep interaction among the game process and the social interaction has been modelled: in this way the player may enforce and discuss the game performances with a wider community of stakeholders. Also the serious game design and construction belongs to the semantic support environment previously discussed: scenarios description terms are disambiguated and addressed to figures and indicators, which may help in choosing and adopting the proper behaviours. These are criteria any kind of training should fulfil regardless of format. Game-learning is able to offer these five characteristics. Game-based training has been reported to offer a safe, effective method of conditioning for people that results in comparable (and, in some cases, greater) improvements in physical and cognitive performance than traditional programs. While technical instruction training has been associated with a higher volume of skill executions (i.e., more ‘touches’), game-based training has been associated with greater cognitive effort, as an important condition for skill learning. Indeed, studies investigating skill learning have reported comparable (and, in some cases, greater) improvements in skill execution, problem solving and decision-making following game-based training rather than training involving repetitious technical instruction. To this end, a RESOLUTE game based meta-application for training will be designed and developed, in order to train different user categories. For instance, people at large can be trained on risk or early warning interpretation, Critical Infrastructure managers on ERMG application, and so forth, according to the learning objectives. Games can produce complex scenarios by simultaneously randomising several conditions and hence help actors learn meta-competences more efficiently.

4.7 Conclusions

With the aim of managing cyber physical systems in critical conditions the European Commission has launched a number of projects to define the European Resilience Management Guidelines, ERMG. Among those projects working on ERMG, RESOLUTE is focussed on the resilience aspects of Urban Transport System, and contribute to the global scenarios creating the guidelines and putting them in place (e.g., operationalize). The followed approach of RESOLUTE is grounded on the possibility of gathering real data from the city which can classified as urban transport system big data, urban data in general, city user behavioural data and social media data. The collection and aggregation of these data sources is a challenge per se, as it is a challenge to exploit them with data mining, data analytics approaches to feed the several decision support systems may be present and are interconnected in the city: mobility, energy, water, telecommunication, services, etc. With this huge complexity on the field, RESOLUTE project aimed at defining guidelines, methods and tools to operationalize the phases of resilience city: preparation, monitoring, reaction, adaptation, learning, and anticipation.

Acknowledgment This work has been supported by the RESOLUTE project (www.RESOLUTE-eu.org) and has been funded within the European Commission's H2020 Programme under contract number 653460. This paper expresses the opinions of the authors and not necessarily those of the European Commission. The European Commission is not liable for any use that may be made of the information contained in this paper. A part of the work described in this paper (namely the Km4City model and tools) has been also supported by Sii-Mobility SCN MIUR project on smart city mobility and transport in Italy.

References

Bellini, E., & Ferreira, P. (2018). Managing interdependencies in critical infrastructures – A cornerstone for system resilience. In *European Safety and Reliability Conference ESREL 2018.*

Bellini, E., & Nesi, P. (2018). Exploiting smart technologies to build Smart Resilient Cities. In P. Gardoni (Ed.), *Smart cities and the role of information and communication technology on achieving sustainability and resilience book – Routledge handbook of sustainable and resilient infrastructure*. Taylor & Francis.

Bellini, P., Benigni, M., Billero, R., Nesi, P., & Rauch, N. (2014). Km4City ontology building vs data harvesting and cleaning for smart-city services. *International Journal of Visual Language and Computing*, Elsevier, 2014. https://doi.org/10.1016/j.jvlc.2014.10.023

Bellini, P., Nesi, P., Simoncini, M., Tibo, A. (2014b). Maintenance and emergency management with an integrated indoor/outdoor navigation support. *International Journal of Visual Language and Computing*, Elsevier, 2014. https://doi.org/10.1016/j.jvlc.2014.10.009

Bellini, P., Cenni, D., & Nesi, P. (2016). AP positioning for estimating people flow as origin destination matrix for smart cities. In *The 22nd international conference on distributed multimedia systems, DMS 2016, Italy, 2016.*

Bellini, E., Nesi, P., Coconea, L., Gaitanidou, E., Ferreira, P., Simões, A., & Candelieri, A. (2016a). Towards resilience operationalization in Urban Transport System: The RESOLUTE project approach. In Walls, Revie & Bedford (Ed.), *Risk, reliability and Safety: Innovation and practice* (European Safety and Reliability Conference (ESREL 2016)). CRC Press.

Bellini, E., Nesi, P., & Ferreira, P. (2016b). *Operationalize Data-driven Resilience in Urban Transport Systems – EPFL- IRGC Guide on resilience 2016.* https://doi.org/10.13140/RG.2.2.12786.09922

Bellini, E., Nesi, P., Pantaleo, G., & Venturi, A. (2016c). Functional resonance analysis method based decision support tool for Urban transport system resilience management. In *IEEE International Smart City Conference (ISC2 2016), Trento.*

Bellini, E., Ceravolo, P., & Nesi, P. (2017). Quantify resilience enhancement of UTS through exploiting connected community and internet of everything emerging technologies. *ACM Transaction on Internet Technology, 18*, 1, Article 7 (October 2017), 34 pages. 27. https://doi.org/10.1145/3137572

Bellini, E., Coconea, L., & Nesi, P. (2019). A functional resonance analysis method driven resilience quantification for socio-technical systems. *IEEE System Journal.*

Dehghani, M. S., Flintsch, G., & McNeil, S. (2014). Impact of road conditions and disruption uncertainties on network vulnerability. *Journal of Infrastructure Systems, 20*, 04014015.

EU. (2012). *Commission staff working document on the review of the European programme for critical infrastructure protection (EPCIP), SWD (2012)190 final.* http://ec.europa.eu/dgs/home-affairs/pdf/policies/crisis_and_terrorism/epcip_swd_2012_190_final.pdf

EU. (2014). *Transport – Connecting Europe's citizens and businesses*, ISBN 978-92-79-42777-0 https://europa.eu/european-union/topics/transport_en

Faturechi, R., & Miller-Hooks, E. (2014). Measuring the performance of transportation infrastructure systems in disaster: A comprehensive review. *Journal of Infrastructure Systems, 21*, 04014025-1.

Ferreira, P., & Simões, A. (2016). *Conceptual Framework, D2.2*, RESOLUTE. http://www.resolute-eu.org/files/D2.2_Conceptual-framework.pdf

Finkelstein, A., & Dowell, J. (1996). A comedy of errors: The London Ambulance service case study. In *Proceeding of 8th international workshop on software specification & design IWSSD-8* (pp. 2–4). IEEE CS Press.

Gaitanidou, E., & Tsami, M. (2016). *ERMG Adaptation to UTS*, RESOLUTE Deliverable 3.7, May 2016 http://www.resolute-eu.org/index.php/deliverables

Gaitanidou, E., Bellini, E., & Ferreira, P. (2016). *European Resilience Management Guidelines*, RESOLUTE Deliverable 3.5, May 2016. http://www.resolute-eu.org/index.php/deliverables

Gehl, P., Wang, M., Taalab, K., D'Ayala, D., Medda, F., et al. (2016). Use of multi-hazard fragility functions for the multi-risk assessment of road networks. In *1st International Conference on Natural Hazards & Infrastructure: INCONHIC 2016, Jun 2016, La Canée, Greece.*

Grasso V., Crisci, A., Nesi, P., Pantaleo, G., Zaza, I., & Gozzini, B., (2016). Public crowd-sensing of heat-waves by social media data. In *16th EMS Annual Meeting & 11th European Conference on Applied Climatology (ECAC), 12–16 September 2016 | Trieste, Italy, CE2/AM3, Delivery and communication of impact based forecasts and risk based warnings.*

Häring, I., Sansavini, G., Bellini, E., Martyn, N., Kovalenko, K., Kitsak, M., Vogelbacher, G., Katharina, R., Bergerhausen, U., Barker, K., & Linkov, I. (2017). Towards a generic resilience management, quantification and development process: General definitions, requirements, methods, techniques and measures, and case studies. In I. Linkov & J. Palma-Oliveira (Eds.), *Resilience and Risk* (NATO Science for Peace and Security Series C: Environmental Security). Springer. https://doi.org/10.1007/978-94-024-1123-2_2

Höller, J., Tsiatsis, V., Mulligan, C., Avesand, S., Karnouskos, S., & Boyle, D. (2014). *From machine-to-machine to the Internet of Things: Introduction to a new age of intelligence.* Academic Press.

Hollnagel, E. (2009). The four cornerstone of resilience engineering. In C. P. Nemeth, E. Hollnagel, & S. Dekker (Eds.), *Resilinece Engeenering Perspective, Vol.2: Preparation and restoration.* Ashgate.

Hollnagel, E. (2011). Prologue: the scope of resilience engineering. In E. Hollnagel, J. Pariès, D. Woods, & J. Wreathall (Eds.), *Resilience engineering in practice – A guidebook* (p. 29-3-39). Ashgate.

IPCC. (2012). Managing the risks of extreme events and disasters to advance climate change adaptation. In C. B. Field, et al. (Eds.), *A special report of Working Groups I and II of the Intergovernmental Panel on Climate Change* (p. 582). Cambridge University Press.

Jenelius, E., & Koutsopoulos, H. N. (2013). Travel time estimation for urban road networks using low frequency probe vehicle data. *Transportation Research Part B: Methodological, 53*, 64–81.

Khademi, N., Balaei, B., Shahri, M., Mirzaei, M., Sarrafi, B., Zahabiun, M., & Mohaymany, A. (2015). Transportion network vulnerability analysis for the case of a catastrophic earthquake. *International Journal of Disaster Risk Reduction, 12*, 234–254.

Larkin, S., Fox-Lent, C., Eisenberg, D., Trump, B., Wallace, S., Chadderton, C., & Linkov, I. (2015). Benchmarking agency and organizational practices in resilience decision making. *Environment Systems and Decisions, 35*(2), 185–195.

Linkov, I., Read, L., Fox-Lent, C., Allen, C. R., Arnott, J. C., Bellini, E., Coaffee, J., Florine, M. V., Hatfield, K., Hynes, W., Jovanovic, A., Kasperson, R., Katzenberger, J., Keys, P. W., Lambert, J. H., Moss, R., Murdoch, P. S., Palma-Oliveira, J., Pulwarty, R. S., … Hyde, I. (2018). Tiered approach to resilience assessment. *Risk Analysis Journal, 38*(9), 1772–1780.

Little, R. (2010). Managing the risk of cascading failure in complex urban infrastructures. In S. In Graham (Ed.), *Disrupted cities – When infrastructure fails* (pp. 27–40). Routledge.

Martelli, C., & Bellini, E.. (2012). The role of the semantic approach to complex statistical information system planning and management. In *IEEE Conference on Signal Image Technology & Internet Based Systems 2012*. Naples.

Mattsson, L. G., & Jenelius, E. (2015). Vulnerability and resilience of transport systems – A discussion of recent research. *Transportation Research Part A, 81*(2015), 16–34.

National Academy of Science (NAS). (2012). http://www.nationalacademies.org/topics/resilience/2012

Navarro, J., Díez, E., Gómez, F., Meneses, R., & Quijano, S. D. (2008). Incertidumbre de las tareas de grupo. Propuesta de un modelo y validación empírica [Group-task uncertainty. Proposal of a model and empirical validation]. *Revista de Psicología Social, 23*(2), 259–273.

Norros, L. (2004). *Acting under uncertainty: The core-task analysis in ecological study of work* (p. 546). VTT Publications. http://www.vtt.fi/inf/pdf/publications/2004/P546.pdf

Reggiani, A. (2013). Network resilience for transport security: Some methodological considerations. *Transport Policy, 28*, 63–68.

Reggiani, A., Nijkamp, P., & Lanzi, D. (2015). Transport resilience and vulnerability: The role of connectivity. *Transportation Research Part A, 81*, 4–15.

RESOLUTE: http://www.resolute-eu.org

UNISDR. (2009). *UNISDR terminology on disaster risk reduction*. United Nations International Strategy for Disaster Reduction. http://www.unisdr.org. Accessed 11 July 2015

United Nations. (2015). *The Sendai Framework, 69th session, Agenda item 19(c)*. This document details The Sendai Framework which was approved by the United Nations in 2015. Retrieved from: http://www.preventionweb.net/files/resolutions/N1516716.pdf

US DOT. (2003). *The public transportation system security and emergency preparedness planning guide, Final Report, January 2003*. https://www.transit.dot.gov/sites/fta.dot.gov/files/docs/PlanningGuide.pdf

Volpe Center (John A. Volpe National Transportation Systems Center). (2013). *Beyond Bouncing Back: Critical transportation infrastructure resilience*. https://www.volpe.dot.gov/events/beyond-bouncing-back-critical-transportation-infrastructure-resilience-conference-package

Woltjer, R. (2006). Resilience assessment based on models of functional resonance. In *Proceedings of the 3rd Symposium on Resilience Engineering*.

Woods, D. D. (2006). Essential characteristics of resilience. In E. Hollnagel, D. D. Woods, & N. Leveson (Eds.), *Resilience engineering: Concepts and precepts* (pp. 21–34). Ashgate.

Chapter 5
Creating a Resilient City: A Community Focused Approach in Bogota, Colombia

Shai Diner

Abstract Resilience is utilised in various areas of academia: physics, economics, urban, individual, community, ecological and social-ecological. When it comes to urban environments, urban resilience often focuses on disaster management without consideration for the community. This article proposes the use of community resilience literature to aid in understanding the role of community within urban resilience. There have been many programs and initiatives, which have enabled cities to begin to address issues relating to resilience; one such city is Bogota. This article puts forward an ethnographic research project focusing on Bogota, Colombia and its relationship with resilience. Observations throughout Bogota were conducted from January to May 2015, supplemented by 15 interviews throughout this period with government officials and various stakeholders. Policy analysis was conducted throughout this period. The results show that through the Bogota Humana Plan, the city of Bogota has been able to draw on various resilience strategies from different academic fields, enhancing their urban, community, and cultural resilience. This article argues that community resilience literature should be utilised within urban resilience strategies, creating a more holistic, cross disciplinary approach which in the long term better for urban environments. The Bogota government provides a quality example of a city, which is implementing resilience strategies utilised within community resilience to establish a holistic resilience urban environment.

Keywords Resilience · Community · Urban · Tools · Auto-ethnography

S. Diner (✉)
RMIT University, Melbourne, VIC, Australia
e-mail: shai.diner@rmit.edu.au

O. F. González Castillo et al. (eds.), *Urban Resilience: Methodologies, Tools and Evaluation*, Resilient Cities, https://doi.org/10.1007/978-3-031-07586-5_5

5.1 Introduction

In the twenty-first century, the majority of the world's population live in urban centres. Given that such high concentrations of the population reside in urban environments, these urban areas must be able to withstand natural and human stressors. Resilience has become central to urban policy, with academics and local governments adopting resilience-based strategies to ensure minimal damage to communities and infrastructure during upheaval.

Bogotá, Colombia has had to overcome multiple obstacles over the last century, providing for a unique context in which to study urban resilience. Colombia had a tumultuous twentieth century, filled with violence and strife (Engel & María, 2013; Bushnell, 1993; Holmes et al., 2007). Bogotá was a victim of these conflicts, which resulted in urban terrorism, inequality, discrimination, class issues and an influx of migration. Over the last two decades, the Bogotá City Government has made concerted efforts to change the urban landscape (Beckett & Godoy, 2009). It therefore serves as an interesting city to study when analysing urban resilience.

Resilience, as a concept has been around for many years however in recent years it has become popular. Resilience has been adopted by various academic disciplines however each discipline utilises the basic meaning of the word; the ability of a biological being, system, structure, person or community, to positively adapt and bounce back despite significant threat or stress (Garmezy, 1991; Waller, 2001; Mcmillan & Chavis, 1986; Godschalk, 2003; Wilkinson, 2011).

This article seeks to discover the strategies the Bogotá city government put in place to successfully enable urban resilience, specifically, focusing on those tactics used to enhance the resiliency of the community as part of the urban structure. A context of Bogota will be provided to understand the specific challenges that Bogota as an urban environment faces. Urban resilience, the ability for an urban environment to bounce back from stressors (human and natural), requires a cross-disciplinary approach when focusing on the human systems within an urban environment and thus, theories utilised in the psychological and community resilience literature will be the focus of the Literature Review. The methodology will go through the auto-ethnographic approach utilised in Bogota, where observations and semi-structured interviews were conducted from January to May 2015. The results section will present the finding of the fieldwork and the discussion will analyse and compare the results with the local government development plan, *Bogotá Humana* (de Bogotá, 2012). This article shares an example of a city that instigated an urban development plan focusing on strengthening its communities and addressing the inequalities to enhance the resiliency of the community, which in turn, aids the ability of the urban environment to bounce back from stressors.

5.1.1 A Colombian Context

Colombia is a rich and contrastive country matched by an equally vibrant culture. Although Colombia had democratic principles in place, conflict has plagued Colombia, during the twentieth century. A central point of conflict has been the diversity of cultural, religious and political beliefs, held throughout the country (Wade, 2000, 2006). Specifically, the class divide existent within post-colonial societies and the varying political ideologies held within those groups has been a source of struggle in Colombia (Safford & Palacios, 2002). The last century in Colombia was a particularly violent period in its history as it was plagued by war throughout the country. The 1900s began on a violent note, with the 1000-day war. This conflict originated between the Liberal and Conservative parties, due to ideological differences. After devastating the countryside, the war ended with the liberal leaders surrendering and the signing of the Treaty of the Wisconsin in 1902 (Bushnell, 1993).

After the 1000-day war, Colombia lived in relative peace until the period now known as *La Violencia*. On April 9th 1948, Jorge Elecier Gaitain, the liberal candidate for the presidential election, was the favourite to win when he was assassinated (Palacious, 2006). This assassination sparked one of the most violent wars in Colombian history lasting for 10 years, between the Liberal Party and the Conservative Party. The war was largely fought in the countryside, resulting in an increase in the migration of rural populations to cities (LaRosa & Mejia, 2012). This internal migration led to a drastic rise in urban populations and was especially felt in Bogotá due to a belief in economic possibilities ("Historia I Portal Bogotá I Bogotá.gov.co", 2015). This population resettlement had a lasting influence on Bogotá. La Violencia ended in 1958 when both political parties agreed to the National Front; an agreement between both parties to alternative governing every four years (Saskiewicz, 2005). Although peace prevailed for 16 years, many social and political problems persisted.

During the National Front there was a rise in political violence as guerrilla groups began establishing themselves around the country, alongside a rise in the drug trade in Colombia. Guerrilla groups with Marxist/ Leninist ideologies, such as the Revolutionary Armed Forces of Colombia (FARC), the National Liberation Army (ELN) and the 19th of April Movement (M-19), began gaining popular support in response to inequalities within the post-colonial country (Henao & Arrubla, 1971). Rural populations were particularly engaged with these groups, as they were often the most affected by the class system and laws relating to land ownership (LaRosa & Mejia, 2012). With popular support amongst these rural populations, guerrilla groups grew in strength throughout the 1970s and 1980s. Right wing paramilitary groups, eventually joined the conflict against the guerrilla groups. At the same time there was a rise in economic violence with marijuana and cocaine being grown, produced and trafficked throughout Colombia. At first, drug cartels controlled this industry, with Pablo Escobar's Medellin cartel the most notorious (Bagley, 1988). M-19, an urban guerrilla group based in Bogotá, became more active as there was growing discord amongst the guerrilla factions. Throughout this period there were bombings on a regular basis in Bogotá. Due to the growing violence in rural areas, migration to Bogotá continued to grow ("Historia I Portal Bogotá I Bogotá.gov.co", 2015). M-19 became increasingly violent in the 1980's, committing some of the most atrocious acts during this period; the most notorious of such acts the Palace of Justice siege (Safford & Palacios, 2002). Eventually, peace talks between M-19 and the Government in 1993 ended the violence occurring throughout Bogotá. While peace was now the norm in this capital city, the urban guerrilla movement and the rapid and extensive internal migration of displaced rural persons had a major, lasting affect on Bogotá and its development.

The District of Bogotá currently consists of 20 municipalities under the jurisdiction of the *Alcalde Mayor de Bogotá DC* (Mayor of Bogota DC). In 1881, the greater metropolitan area had a population of around 84,000 people; in 1951 it had grown to 700,000 people; in 1964 to 1,600,000 people; in 1973 2,500,000 lived in Bogotá; in 1985, 4100,00; in 1993, reaching almost 6,000,000 people. The current population is almost 8 million inhabitants. These figures demonstrate the rapid urbanisation of Bogotá, which has been critical to the current urban landscape (Historia ("Historia I Portal Bogotá I Bogotá.gov.co," 2015). Bogotá's homicide rate went down considerably between 2005 and 2012, fluctuating from 24 to 16.5 persons per 100,000 people (UNOCD, 2013). Although this figure fluctuated considerably, it has lowered substantially, alongside, the national average.

Due to Bogotá's recent history, the people and communities of Bogotá have had to overcome immense suffering and violence, alongside rapid urbanisation. These difficulties are still existent within Bogotá, with many social issues influencing locals' lives. Inequality and discrimination is still strife for young people, women, LGBTI and displaced persons and ethnic and racial communities (Lampis, 2013). Furthermore, lack of infrastructure means travel within the city impinges on many people's lives, with individuals having to travel for three to four hours per day between work and home. As a result of internal migration, particular areas of the city are underdeveloped as the class divide widens within Colombia, creating further hardship for the urban poor (Programa de Gobierno I Portal Bogotá I Bogota.gov.co, 2015). Although Bogotá lives in relative peace, vulnerable populations feel the effects of the class divide, poor infrastructure, and migration.

The Lesbian, Gay, Bi-sexual, Transgender and Intersex (LGBTI) population of Bogota, is one such vulnerable community. Although the LGBTI community have obtained certain rights, through a series of high court rulings, LGBTI people still face discrimination, recrimination and rejection from their families and communities (Ripoll, 2009; Sotto & Moreno, 2012). Within public spheres, LGBTI people are in fear of violence. There is also a small population of people who are especially vulnerable due to their status as, 'internally displaced, homosexual / transgender and economically disadvantaged', which is a group of people who often go unseen (Zea et al., 2014, p. 14; Bianchi et al., 2014).

Over the years, the government has created certain policies to aid the city to overcome its violent history and current urban concerns. Since 1995, the Bogotá government has become more active in transforming the urban landscape (Montezuma, 2005; Heres et al., 2014).

Bogotá, with its history of violence and internal migration, resulting in its current urban concerns provides for an interesting city to analyse how a local government is utilising cultural and community policy to overcome the current inequalities that it faces. This paper argues that the Bogotá government, within its urban resilience strategy, *Bogotá Humana*, draws upon culture and community in providing the tools to build individuals' and communities' resilience, to overcome its violent past and to cope with future disasters.

5.2 Literature Review

Resilience has been utilised in physics, ecology, psychology, economics and the social sciences (urban, community, social). This section will review the relevant literature to understand resilience in an urban environment, specifically examining urban, psychological and community resilience.

5.2.1 *Urban Resilience*

Urban resilience has been conceptualised in a variety of manners: as a metaphor, as disaster management and as urban hazard mitigation. However, In order to coneptualise urban centres, the urban enviornment must be understood as a complex and interdependent system, with various vulnerabilities demanding consideration. The urban enviornment can be understood as a human ecosysytem. The Human Ecosystem Model (Fig. 5.1), is a conceptual framework that demonstrates the linkages between the human and the ecological systems and their interdependency (Pickett et al., 1997).

Human Ecosystem Model is a conceptual tool to understand the linkages between the human and natural elements. The model is divided into two sections: the human social system and a resource system which compromises elements of human resources and ecosystems. The resource system and specifcally, the ecosystem is the foundation for the human system to exist. This model also indicates that cultural and socio economic resources are part of the foundation of the human social system and are required for it to function (Pickett et al., 1997; Machlis et al., 2008).

It is with this understanding that urban resilience must be considered. Because the social-ecological system within an urban enviornment is unique, the city must consider both the ecology and the humans to ensure that the overall system functions well.

David Godschalk defined a resilient city as a "a sustainable network of physical systems and human communities" (Godschalk, 2003, p. 17). Godschalk describes the physical systems and environmental components, (natural and built), which include roads, infrastructure, communication, sewerage, waterways and the

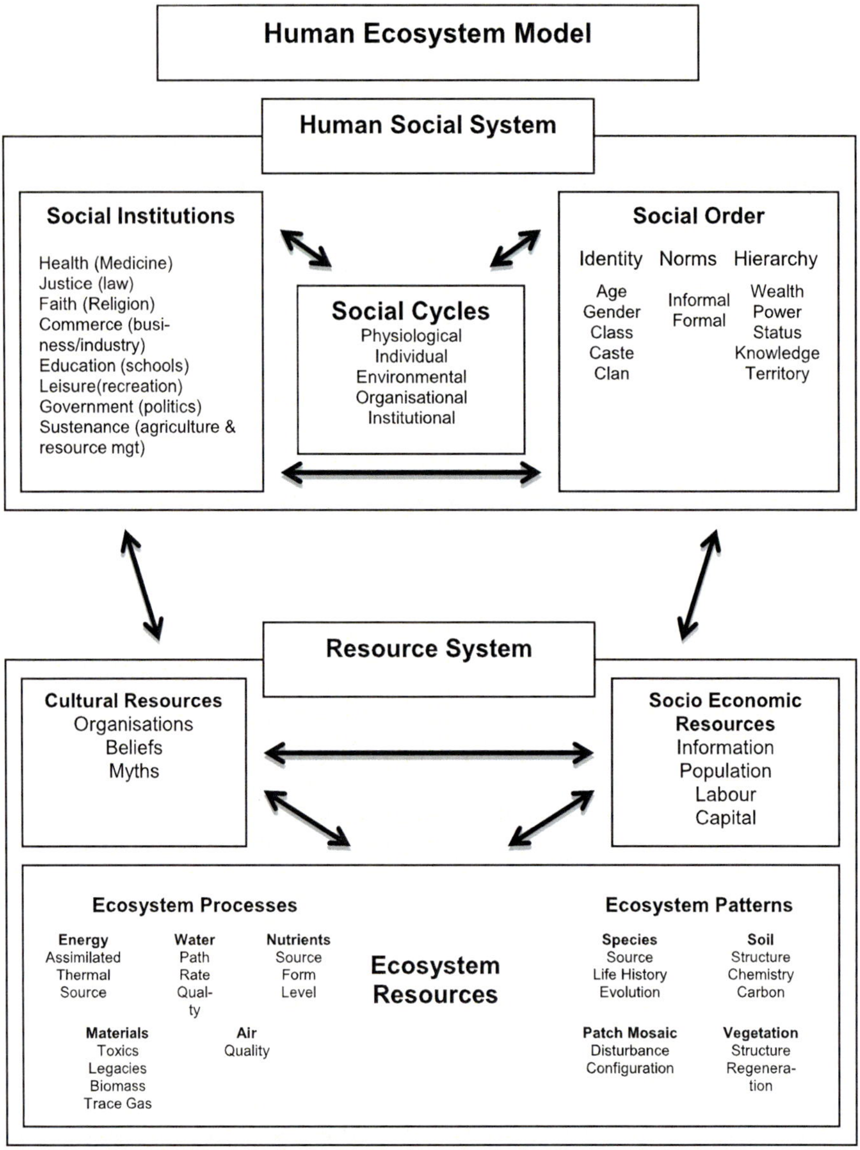

Fig. 5.1 Human Ecosystem Model demonstrates the linkages between the human social system, and the resources which humans draw upon (cultural and socio economic and ecosystem resources). (Source: Pickett et al., 1997)

ecological elements of a city. Human communities are the social institutions, which constitute a city, including schools, neighborhoods, organizations and social services.

Disaster management is a key concept of urban resilience, focuses on environmental factors concerning a city. Disaster management is a set of strategies, different for each city, implemented to manage potential disasters effectively

(Comfort, 1988; Boin & Mcconnell, 2007). These plans are aimed at managing disasters such as floods, fires, earhtquakes, hurricanes, diseases, warfare, terrorism, riots and industrial action (Vale & Campanella, 2005). Disaster management strategies might include: evacuation plans, proper drainage (in flood or heavy rain areas), air traffic diversion plans and emergency shelter plans (Vale & Campanella, 2005). These strategies are of exceptional import to allow urban areas to recover from disasters. However, each city has its own hazards and disasters it must prepare for, depending on its natural environment, infrastructure and socio-political climate. Given this individuality of urban areas, there is no 'one size fits all' approach.

Urban resilience discussion often focuses primarily on disaster mangement with a heavy emphasis on infrastructure and planning. And while various authors have provided academic theories incorporating the social compononent, there is a dearth of literature on the practical aspect of increasing the community component of resilience within an urban environment (Wilkinson, 2011; Vale & Campanella, 2005; Adger et al., 2005; Crane, 2010; Gunderson, 2010; Pretty, 2011).

This article looks to fill that gap by focusing on the human component of urban enviornments and explore how government can aid in building the necessary tools to aid a communities ability to be resilient.

This paper proposes that successful increasing of urban resilience requires an incoporation of all resilience disciplines. Urban resilience, which is often focused on the physical environment, requires a greater understanding of the human system and it's impact on a city's resilience. A review of the literature on individual and community resilience provides an understanding of the human component of the human-ecosystem.

5.2.2 *Individual Resilience*

Individual resilience, also known as psychological resilience, refers to the ability of a person to overcome and 'bounce back' from destructive pressures that they encounter.

In its most simplistic form, this type of resilience is defined as an indvidual's ability to properly recover from challenging life events (Garmezy, 1991). This can be further defined as one's ability or capacity to successfully adapt to extenuating stressful circumstances (Masten et al., 1990). This can be advanced by exploiring how such resilience develops, and defines resilience as, the ability to withstand and cope with adverse situations, which are influenced by risk factors and protective and risk mechanisms (Rutter, 1985).

Protective factors are tools that support one's resiliency. These protective mechanisms have been categorized into several broad areas: individual support, family support and societal support (Olsson et al., 2003; Rutter, 2012). Each support apparatus provides tools for individuals to draw on in times of crises. These mechanisms are learnt behaviours, which are developed over the course of a lifetime. Therefore individuals have the ability, through life experiences, to become more resilient.

Individual protective mechanisms include a person's genetics, their sociability and personal attributes (Smith, 1999; Luthar & Zigler, 1991; Luthar, 1991; Allen, 1998). Family support mechanisms are parental love, strong family unit and being valued by others (Olsson et al., 2003; Smith, 1999; Egeland et al., 1993). Societal supports include a person's socioeconomic status, schools, communities, cultural beliefs and readily available resources (Rutter, 1987; Wolff, 1996; Werner, 1995). Another component of societal support is social connectedness. This is a sense of community, a factor that evidence demonstrates to be very important to well-being, and a key tool in overcoming hardships (Mcmillan & Chavis, 1986; Sonn & Fisher, 1988). These support mechanisms greatly aid an individual's ability to withstand and overcome difficult circumstances. To best discern what strategies are being utilised successfully in the urban resilience development plan, *Bogotá Humana* understanding how individuals cope with stressors is imperative.

Determining an individual's resilience has been the subject of extensive research and multiple measurement models (Connor & Davidson, 2003; Antonovsky, 1979, 1987; Smith et al., 2008). However, this paper is not concerned with determining individual's resilience; rather its aim is to understand if and how a government can aid in building resilience.

5.2.3 *Community Resilience*

Due to the strong influence that communities have on their citizens, extensive research has been conducted into community and cultural resilience. Although the concept of community is explored within the construct of psychological resilience, it has developed into its own category due to the growing understanding of the importance of a community's ability to withstand shock.

Community resilience, can be described as "the capability of a system to maintain its functions and structure in the face of internal and external change and to degrade gracefully when it must" (Allenby & Fink, 2005, p. 1034) Coles and Buckle take a more utilitarian stance, defining it as the community's ability and proficiency to participate in disaster recovery (Coles & Buckle, 2004). Yet the explanation of resilience, as "the ability of community members to take meaningful, deliberate, collective action to remedy the effect of a problem, including the ability to interpret the environment, intervene and move on" seems most appropriate for the research presented here, as it focuses on action taken by the community to overcome stressors (Pfefferbaum et al., 2007, p. 349). This study is interested in the action taken by the government- the leaders, as it were, of the community- to provide the tools to overcome stressors.

Like individual resilience, community resilience provides resources to aid in the community's recovery. Community resilience has been conceptualised as a metaphor and developed into a framework, which aids in understanding the various components which build community resilience (Norris et al., 2008). As seen in Fig. 5.2, community resilience resources compromise: information and community;

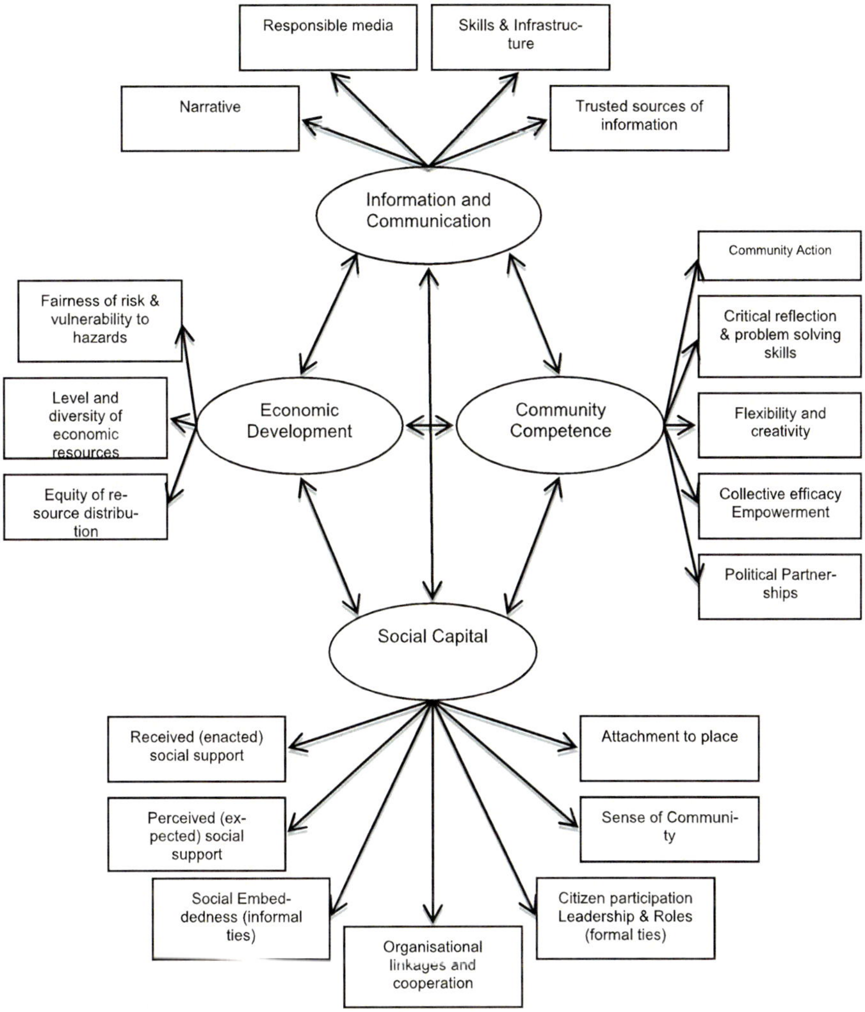

Fig. 5.2 Community Resilience Framework (Community resilience a set of metaphors). (Source: Norris et al., 2008)

economic development; social capital; and community competence. Economic development is about a community having the appropriate resources, physical capital, employment opportunities, health services and schools. Having economic resources greatly builds the foundations for community resilience (Godschalk, 2003; Rose, 1999, 2005; Cutter et al., 2003). Social capital is a contested concept, which has been debated by various authors. For the purposes of this study however, social capital refers to investment into social networks, providing benefits and resources to those communities and individuals. This is achieved through social support, formal and informal ties within communities and a sense of community

(Pretty, 2011; Morrow, 1999). Information and communication relates to the requirement to provide accurate information to the public and ensuring that communication is collaborative, which will aid in the communities resilience (Pfefferbaum et al., 2007; Longstaff et al., 2010; Goodman et al., 1998). Community competence pertains to the ability of a community to function as a unit, a cohesive network of people, through collective action, decision making, critical reflection, leadership and empowerment (Goodman et al., 1998; Brown & Kulig, 1996). For these resources to be effective, they need the properties of robustness; redundancy; rapidity; and resourcefulness (Norris et al., 2008).

If a community has these resources, it will have the ability to overcome challenging circumstances, regardless of the details of the catastrophe (Coles & Buckle, 2004). Utilising community resilience framework, this article will work to understand the relationship between the strategies that government enact and the tools required for urban and community resilience.

A variety of different tools have been developed to determine how resilient communities are however each analytical model has its own limitations and focus, with the majority of tools more closely observing disaster management than the community's rebound (Cutter et al., 2003; Kulig et al., 2013, 2008a, b; Brodsky et al., 1999; Bruneau et al., 2003). Although the bricks and mortar of disaster recovery is relevant to community resilience, this present study is not concerned with discerning whether the community is itself resilient. Rather, it is analysing the strategies and tools that have been created and enacted by an urban government to promote culture and a sense of community that aid in the communities ability to be resilient.

The literature on individual and community resilience provides the framework to understand how the human systems within an urban area can recover from external stressors. Specifically, the sense of community and social-connectedness, which aids in one's ability to overcome stressors, will be analysed to understand how a government is enhancing their urban resilience.

Bogotá provides an instructive example to demonstrate how a city is utilising culture to promote peace and build community, and in doing so, providing the necessary tools to increase resilence of both people and neighborhood. Resilience, as a learned trait, allows for people to be receptive to culture and community building.

5.3 Methodology

This project utilises an auto-ethnographic approach comprised of observations and interviews alongside a policy analysis in order understand how the Bogotá government enhances individual and community resilience within an urban framework (Mcilveen, 2008; Ellis et al., 2000, 2011; Wall, 2008). Initial fieldwork was conducted between January – May 2015. Auto-ethnographic observations capture researcher impressions of public venues and events in Bogotá, which is documented through extensive field notes (Fetterman, 2010; Hammersley & Atkinson, 2007; Lüders, 2004). Throughout the field notes, the researcher utilised reflexivity, acknowledging their presence as a research and as an outsider (Ellis & Bochne, 2000). This is apparent in the results, where the researcher utilises first person to describe events that occurred in the field. 15 stakeholders in Bogotá participated in a series of semi-structured interviews (Holstein & Gubrium, 1999; Hopf, 2004). Subjects included, advocacy workers, academics, community workers and government workers and officials. Questions ranged from general queries about Bogotá to the functions and scope of the local government to specifics about social programs. Interviews ran between 1 and 2 h, with the majority of the interviews conducted in Spanish. These were transcribed and translated by the researcher. Two interviews were conducted in English and transcribed by the researcher (Britten, 2006).

The policy document analysed for the purposes of this study was 'Plan de Desarrollo, 2012–2016, *Bogotá Humana*" (Development Plan, 2012–2016, Human Bogotá)", which came into effect mid year, 2012. This policy was developed under the Mayor, Gustavo Pedro. Policy analysis was conducted after the fieldwork, with a study of the *Bogotá Humana* Plan as well as the strategies put in place to enable cultural and community resilience.

Patterns and themes that became apparent throughout fieldwork, and in analysis of the interviews, observations and policy, were coded into categories (Fetterman, 2010). One common theme that appeared throughout the fieldwork, was the government's active role with the various communities and municipalities within Bogotá.

There were a few limitations to this research, which should be noted. As a non-native speaker with intermediate verbal Spanish skills, it took the researcher 2 months to be able to communicate easily in Spanish. Locals were understanding of the researcher's Spanish and were inclusive and considerate with my level of Spanish. As time passed and the researcher's Spanish grew, closer bonds were formed with the local experts assisting the research. This allowed for further access to various communities.

5.4 Results

After a couple weeks in Bogotá, I began noticing, the *Bogotá Humana* (Human Bogotá) logo everywhere: electrical workers on *Carrera 19*, road workers on *Avenida Caracas*, street performers on *La Septima*. Having already read parts of the policy, I had not imagined its scope and reach.

Bogotá Humana, is a 489 page, urban development plan, which outlines strategies for the development of Bogotá over a 5 year period. The general objective provides an insight into the ideology behind the plan:

> The Human Development Plan Bogota has the overall aim to improve human development in the city, giving priority to children and adolescents with emphasis on early childhood and applying a differential approach in all its policies. Bogota seeks to address forms of social segregation, economic, and cultural space, by increasing the capacities of the population on the full enjoyment of rights, equal access to the enjoyment of the city, reduce support development of national economy, as well as seek to ease the burden of the expense of the poor-

> est and defence policy advocacy and protection of human rights of citizens. Also, the Plan will contribute to land management around water, minimizing future vulnerabilities arising from climate change and protecting the main priority in ecological structure of the city, as the basis for a new model of urban growth based on environmental sustainability, including the urgent revitalization of rural areas and democratic expression of land use, and promotion of a multimodal transport system.

The general aim of *Bogotá Humana* demonstrates the ideology and driving force behind the policy document. It is aiming to curb social, cultural and economic inequalities throughout the city, reduce human rights violations whilst also tackling the environmental concerns that Bogotá faces in light of climate change. The policy outlines environmental, infrastructure and human system strategies, with the ultimate focus on improving human life for the citizens of Bogotá.

The policy will be explored by comparing the policy document to the lived experience of the researcher and their field notes. This will be done utilising two examples: LGBTI Centres and IDARTES.

5.4.1 LGBTI Centres

Article 11 in *Bogata Humana* is dedicated to fighting all forms of discrimination and violence, including that of the LGBTI population. The goals of article 11 are to reduce in Bogata: discrimination by 20%; child labor to less the 1.5%; and teenage pregnancy by 16%. Furthermore, the plan aims "to reduce by 15 per cent the perception of discrimination, violence and social exclusion of people in the LGBTI sector".

Subdirección para Asuntos LGBT (Sub-department for LGBT Affairs) is part of the *Secretaria Distrital de Integración Social* (District Secretariat for Social Integration) a government department consisting of approximately 350 people doing a variety of work, including, running two communities centres. Although the department only mentions LGBTI, it provides resources for the entire community, including queer and questioning individuals.

Centro Comunitatio Distrital LGBTI (LGBTI District Community Centres) have existed in Bogotá prior to this policy, and were established in 2006 (Serrano-Amaya, 2011). *Bogotá Humana* continues with the model of LGBTI centres funding two such community centres, which were operating during my fieldwork; one in the suburb of Tuesaquillo and the other in Santa Fe.

I had the good fortune to be introduced to some of the people working at the LGBTI community centres in Bogotá. I spent some time at both centres to gain an understanding of what goes on in this city and what the centres provide. Over time, I got to know some of the staff and regular attendees and felt welcomed, particularly towards the end. The more I time I spent there, the more I realised that the centre was, quite literally, a community. It was a place for persons who identify as LGBTI, to attend and be part of a community.

Bogotá is divided into different stratums – ranging from 0 to 6, based upon each area's facilities and infrastructure. Stratification is thus an indication of the suburb's socioeconomic level, meaning the higher the stratification, the wealthier the suburb (Gaviria et al., 2010; Medina et al., 2008). The Tuesaquillo centre is located in a suburb with a stratum of 4, whilst the Santa Fe suburb is 2. Thus, the clientele of each centre are different and the purpose changes based on the population attending. For example, the Santa Fe centre focuses more on fundamental necessities (e.g. education, employment and health) whilst the Tuesaquillo centre has a greater emphasis on support groups and cultural events.

Many attending the Santa Fe centre engage in prostitution and drugs, often living on the streets with little to no social support, as often, they are internally displaced persons due to rural violence (Zea et al., 2014). The centre provides a safe environment for a vulnerable population of sex workers and drug users. The Santa Fe centre run workshops, provides meals on a daily basis, runs education classes and provides information on safe sex practises.

The Tuesaquillo centre delivers programs for the entire LGBTI community. There are groups for a diverse range of people, with a variety of activities; dance groups, psychology groups, English groups and trans* groups, providing support to individuals who attend these groups.

There was one particular day where I attended a trans* meeting at the Tuesaquillo centre. I spent 3 h in the one room, full of people I had never met, and by the end of the 3 h, I felt a part of a community group. This group of trans* women gather every week to spend time with each other and get to know other people who are similar to them. There was such a strong sense of comradeship and community with these people, even though I had just met them. We spent 3 h talking about music interests and making decoration for a dance coming up in 3 weeks. And when things were winding down and I was getting ready to go home, I got asked to join them for bread and soda. And so I tagged along to their weekly tradition of breaking bread and drinking a cup of soft drink. As we sit in this bakery, which is otherwise empty, the room seems filled with people as they begin to get to know, Shai, the Australian and I get to know them. Their honesty and openness about themselves and their lives was confronting but exceptionally warming. I instantly felt the sense of community that these individuals feel.

5.4.2 *Cultural Freedom, IDARTES and a Festival*

Article 14 of the *Bogotá Humana*, "Cultural Freedom and Sport Program" outlines strategies to recognize and promote the dimensions of art and culture, and physical activity, recreation and sport. "The program promotes the recognition and the deployment of the artistic, cultural, recreational and sporting practices of different populations, reducing the economic, social, territorial barriers and mediating before the cultural barriers that limit their free exercise and visibility".

The policy has specific and ambitious goals, with the intention to increase cultural and sporting organisations by 60%; generate 5000 jobs; and increase the reach of the cultural, recreational and sport services by 35%. This section will use one example to demonstrate how this goal is being achieved, although by no means is this the only way. Throughout the fieldwork period, there were extensive examples that demonstrate the effectiveness of this policy. One such example is the promotions of arts and culture by "Instituto Distrital de Las Artes" (District Institute of the Arts) (IDARTES).

IDARTES is a governmental department, established four years ago, as part of the "*Secretaría de Cultura, Recreación y Deporte*" (Secretariat of Culture, Recreation and Sport). IDARTES *is* solely dedicated to the promotion of arts and culture throughout the city, organising some of the main cultural events that occur in Bogotá. IDARTES has 6 departments: Music, Theatre, Literature, Dance, Visual and Plastic Art and Audio-visual Arts. IDARTES works throughout the city, among all strata, promoting arts and culture: from peace marches to the biggest rock concert in Latin America.

One way in which IDARTES interacts with the community, is through open discussions, as this interviewee states: "*Mesas sectoriales*", which means a place where we invite people to collaborate and do projects together." The interviewee then elaborates about *Mesa sectoriales* within the dance sector, a collaborative process where IDARTES works with local artists to create and produce works. Where possible, IDARTES provide the appropriate support. The process of the government and artists working together is incredibly important not only for artists within the industry, but also to allow for cultural reach to exceed its boundaries by enabling more artistic work to be produced. This can be seen in a festival that was run by the government, in collaboration with artists, theatre and performers.

> Another way is creating shows, creating events with topics that we know that the city is thinking about, like Cumbre, por las artists, a way to interact with the community. the general community of Bogotá.

5.4.2.1 La Cumbre

In the beginning of April, two different people told me about an event coming up in Bogotá: "*Cumbre Mundial de Arte & Cultura Para La Paz* de Colombia "(World Summit of Art and Culture for Peace in Colombia). Little did I realise that a dance concert I had been planning to attend for some time was actually a part of a much bigger event.

The *Cumbre* is a week long festival run by the local government, focusing on four pillars: Art and Culture in Peace Building; Memory, Stories, and Communication; Territories and Culture of Peace; Cultural and Symbolic Reparation. These pillars, all related to conflict and resolution, were created to help attendees better understand the "processes of building peace with social justice, repair processes and the creation of conditions of non-repetition" (Instituto Distrital de las Artes, *Idartes, gov.co*, 2015, p. 3). The *Cumbre* consisted of a multitude of offerings: academic panels, Q and A's, music and dance concerts, a peace march and theatrical performances. The festival ran throughout the city at different locations, depending upon the event, with free admission to everything. Around 300 international, national and local speakers and participants were invited to partake in the festival. As I rummaged through the program, the list of events I wanted to attend grew exponentially, along with my excitement. For the purposes of this paper, I will not speak about all the events, rather I will focus on two superb events which demonstrate the promotion of peace through culture.

5.4.2.2 Hip-Hop

I decided to attend a conference day at Jorge Elecier Gaitin theatre. The theme for this afternoon was hip-hop; with three different panel session. The first session was with academics, the second with Colombian hip hop artist. The final session, occurred at nighttime, with Zulu Nation and African Bambata, was conducted in English with a Spanish translator. This was the busiest session of the day, with every seat in the theatre taken and isles filled with people. This last session was interesting and thrilling at the same time. Although the questions became repetitive after awhile, with many questions revolving around the use of hip hop in Bogota, the crowd took control of the event with many women speaking up about the need for diversity and greater representation of women in hip-hop. I believe that one of the most striking aspects of this event, and of hip hop culture, specifically that of Zulu nation, is there promotion of peace and love, rather than violence. The speakers kept emphasising that work, love, peace and fun are the pillars of hip-hop, not violence. Every time this was mentioned the crowd shouted and screamed in support. There was a feeling and sense of community in the theatre. It felt that everyone was there with a purpose. They all loved the hip-hop movement and believed in its ability to change lives, as it had changed many of their lives. This is a feeling I will not forget. Many of the lower socio economic communities have become part of the hip hop movement, with the government encouraging them along the way. Through dance programs run with young people, too welcoming graffiti art into the city and providing a visually stunning landscape for its citizens. The sense of comradeship in the theatre on this night, was vibrant. As an outsider to this community, I never thought that I could feel a part of it, and for this brief few hours, I could feel the sense of community that was existent within the hip hop community in Bogota.

5.4.2.3 March for Peace

As part of the festival there was a March for Peace (*Movilización April 9*) organised on April 9th. April 9th, as mentioned previous, is a significant day for Colombians, due to the assassination of Jorge Eleicier Gaitin on that day in 1948 as well as being the National Day of Remembrance and Solidarity for Victims. I didn't know too much about the march, but I did know that at the end of the march, there would be a concert. I also didn't have high expectations as it was a Thursday but I was blown away by the huge attendance of people. For a Thursday afternoon there were a considerable amount of people marching for peace. Majority of the people walking were wearing white, however there were groups in matching colours and shirts; groups on bikes; youth groups; political groups. There were a variety of different banners and posters, all with a message about peace. There were a variety of different flags representing different ethic, cultural, political, religious and social groups. The diversity amongst the people and groups shocked and inspired me. All these different people, from all different walks of life, were all chanting in support of peace in their city and country.

As an outsider to this city, hobbling through the march by myself, I did not feel very connected to the cause nor did I feel part of something bigger. This cause, although is something I believe in, is not my pain. I did not go through the hardships, nor was I affected by this conflict. So although I can support it, I felt very much the outsider.

At the end of the march there was a concert with *Orquesta Filarmonic Juvenil* and a choir of 450 children. Famous Colombian and Latin American singers and artists accompanied them. When the first song started, "*Solo le pido a Dios*" (Just ask God), this seemingly disinterested crowd began to get involved. It felt like something really special. It felt as if the whole crowd were swaying and singing together, and were coming together through music. I did not know the song, and therefore could not feel a part of the crowd, but I did appreciate the vibrancy off it.

There was a feeling at this March that the locals want to move on from its conflict ridden past and begin a new peaceful era. This could be seen through the sheer numbers who attended the march on a Thursday afternoon. This march was also very significant in Colombia at the time, because there were peace talks occurring between the federal government and FARC in Cuba, which had been the longest peace negotiations in the countries history.

5.5 Discussion

The choice of the name, *Bogotá Humana*, was a very deliberate choice, it seems. This plan was focused around the humans within Bogotá, and ensuring that their quality of life improved. Even the ecological elements of this policy, were driven with an underlying theme for the humans, to ensure their safety and well being.

The policy specified that there would be an LGBTI centre however it did not detail the extent to which centres would operate. These two LGBTI centres, provide a space for a community to form, which appears to have occurred in these centres. Various communities have formed around these centres, with young gay male networks and a trans* group. These centres provide a safe environment for a vulnerable population, allowing individuals to meet like-minded people and express their true self without fear of recrimination. The sense of community felt within these centres was evident from observations and interviews. The community centres function as social capital providing social support, formal and informal ties to a greater social structure and a sense of community. The government is providing a variety support mechanisms for a vulnerable population in Bogotá. This is allows individuals and the communities surrounding the centres the mechanisms and tools to aid in their resilience, contributing to their community's resiliency. This then directly affects those individuals and the communities around them to enhance their urban environments.

IDARTES as a cultural organisation is an exceptionally powerful tool that the city can utilise to promote its culture and to create social change. IDARTES provide an avenue for people to express culture and to provide a sense of community. With events such The *Cumbre*, allow for people to feel a part of a bigger community, to know that they are not alone. The inclusivity of the events, due to its freeness and accessibility, speaks volumes to the commitment of the government to ensuring cultural events are accessible to all citizens of Bogotá, regardless of class, ethnicity or race. Furthermore, the collaboration process by which IDARTES follows is important to ensuring that artists have the resources and space to promote culture. IDARTES functions as social capital and community competence, within resilience framework, by providing the organisation linkages and cooperation, leadership, political partnerships, empowerment, flexibility and a sense of community.

Although The Cumbre is only one festival, it demonstrates the use of culture to promote peace. The mass march allowed for people who have never encountered before, from all different social classes, to feel a sense of community. The concert which followed, brought people together through music, creating a shared experience. The hip-hop events empowered a vulnerable population by providing many free events for the community to attend. The *Cumbre*, functions as a social capital and community competence on many levels. It provides a sense of community, a community of peace, allowed for community action, demonstrated political partnerships, provided social ties and support and most of all, empowered the citizens of Bogotá. With Bogotás violent history, the promotion of peace is highly important by the government, but the creation of community surrounding peace, provides the tools the citizens require to form a resilience community.

5.6 Conclusion

This article has examined literature pertaining to urban, psychological and community resilience, arguing that in order to create a truly resilient city, community resilience should be adopted within urban resilience policy. The article provided an example of a city, Bogotá that adopted aspects of community resilience as a way for the city to bounce back from it violent past. The *Bogotá Humana* development plan provides the tools to increase the ability of the community to be resilient however further research must be conducted to determine whether the policy as a whole is effective.

References

Adger, W. N., Hughes, T. P., Folke, C., Carpenter, S. R., & Rockström, J. (2005). Social and ecological resilience: Are they related ? *Science (80-.), 309*(5373), 1036–1039.

Allen, J. R. (1998). Of resilience, vulnerability, and a woman who never lived. *Child and Adolescent Psychiatric Clinics of North America, 7*, 53–71.

Allenby, B., & Fink, J. (2005). Toward inherently secure and resilient societies. *Science (80), 309*(5737), 1034–1036.

Antonovsky, A. (1979). *Health, stress, and coping*. Jossey-Bass Publishers.

Antonovsky, A. (1987). *Nraveling the mystery of health: How people manage stress and stay well.* Jossey-Bass.

Bagley, B. M. (1988). Colombia and the war on drugs. *Foreign Affairs, 67*(1), 70–92.

Beckett, K., & Godoy, A. (2009). A tale of two cities: A comparative analysis of quality of life initiatives in New York and Bogota. *Urban Studies, 47*(2), 277–301.

Bianchi, F. T., Reisen, C. A., Zea, M. C., Vidal-Ortiz, S., Gonzales, F. A., Betancourt, F., Aguilar, M., & Poppen, P. J. (2014). Sex work among men who have sex with men and transgender women in Bogotá. *Archives of Sexual Behavior, 43*(8), 1637–1650.

Boin, A., & Mcconnell, A. (2007). Preparing for critical infrastructure breakdowns: The limits of crisis management and the need for resilience. *Journal of Contingencies & Crisis Management, 15*(1), 50–59.

Britten, N. (2006). Qualitative interviews. In C. Pope 3rd & N. Mays (Eds.), *Qualitative research in health care* (pp. 12–20). Blackwell Publishing Ltd.

Brodsky, A. E., O'Campo, P. J., & Aronson, R. E. (1999). PSOC in community context: Multi-level correlates of a measure of psychological sense of community in low-Income, urban neighborhoods. *Journal of Community Psychology, 27*(6), 659–679.

Brown, D., & Kulig, J. (1996). The concepts of resiliency: Theoretical lessons from community research. *Health and Canadian Society, 4*(1), 29–52.

Bruneau, M., Chang, S. E., Eguchi, R. T., Lee, G. C., O'Rourke, T. D., Reinhorn, A. M., Shinozuka, M., Tierney, K., Wallace, W. A., & Von Winterfeldt, D. (2003). A framework to quantitatively assess and enhance the seismic resilience of communities. *Earthquake Spectra, 19*, 733.

Bushnell, D. (1993). *The making of modern Colombia: A nation in spite of itself.* University of California Press.

Coles, E., & Buckle, P. (2004). Developing community resilience as a foundation for effective disaster recovery. *Australian Journal of Emergency Management, 19*, 6–15.

Comfort, L. K. (1988). *Managing disaster: Strategies and policy perspectives.* Duke University Press.

Connor, K. M., & Davidson, J. R. T. (2003). Development of a new resiliene scale: The Connor-Davidson resilience scale (CD- RISC). *Depression and Anxiety, 82*, 76–82.

Crane, T. (2010). Of models and meanings: Cultural resilience in social – Ecological systems. *Ecology and Society, 15*(4), 19.

Cutter, S. L., Carolina, S., Boruff, B. J., Carolina, S., Shirley, W. L., & Carolina, S. (2003). Social vulnerability to environmental hazards. *Social Science Quarterly, 84*(2), 242–261.

de Bogotá, A. M. (2012). *Plan de desarrollo económico y social y de obras públicas para Bogotá distrito capital 2012–2016: Bogotá humana.*

Egeland, B., Carlson, E., & Sroufe, L. A. (1993). Resilience as process. *Development and Psychopathology, 5*, 517–528.

Ellis, C. S. & Bochne, A. (2000). Autoethnography, personal narrative, reflexivity: Researcher as subject.

Ellis, C., Adams, T. E., & Bochner, A. P. (2011). Autoethnography: An overview. *Historical Social Research, 36*(4), 273–290.

Engel, S., & María, A. (2013). Displacement due to violence in Colombia: A household-level analysis. *Economic Development and Cultural Change, 55*(2), 335–365.

Fetterman, D. M. (2010). *Ethnography Step By Step* (3rd ed.). Sage.

Garmezy, N. (1991). Resilience in children's adaptation to negative life events and stressed environment. *Pediatric Annals, 20*, 459–466.

Gaviria, A., Medina, C., & Morales, L. (2010). The cost of avoiding crime the case of Bogotá. In *The economics of crime: Lessons for and from Latin America* (pp. 101–132). University of Chicago Press.

Godschalk, D. R. (2003). Urban hazard mitigation: Creating resilient cities. *Natural Hazards Review, 4*(3), 136–143.

Goodman, R. M., Speers, M. A., Mcleroy, K., Fawcett, S., Kegler, M., Parker, E., Smith, S. R., Sterling, T. D., & Wallerstein, N. (1998). Identifying and defining the dimensions of community capacity to provide a basis for measurement. *Health Education & Behavior, 25*(3), 258–278.

Gunderson, L. (2010). Ecological and human community resilience in response to natural. *Ecology and Society, 15*(2), 8.

Hammersley, M., & Atkinson, P. (2007). *Ethnography: Principles in practice* (Vol. 10, no. 4). Routledge.

Henao, J. M., & Arrubla, G. (1971). *History of Colombia*. Kennikat Press.

Heres, D. R., Jack, D., & Salon, D. (2014). Do public transport investments promote urban economic development? Evidence from bus rapid transit in Bogota. *Transportation (Amst), 41*, 57–74.

Historia | Portal Bogota | Bogota.gov.co, 2015. [Online]. Available: http://www.bogota.gov.co/ciudad/historia. Accessed 27 Sept 2015.

Holmes, J. S., De Piñeres, S. A. G., & Curtin, K. M. (2007). A subnational study of insurgency: FARC violence in the 1990s. *Studies in Conflict & Terrorism, 30*(3), 249–265.

Holstein, J. A., & Gubrium, J. F. (1999). Active interviewing. In A. Bryman & R. G. Burgess (Eds.), *Qualitative research* (Vol. 2, pp. 105–121). Sage.

Hopf, C. (2004). Qualitative interviews: An overview. In U. Flick, K. von Ernst, & I. Steinke (Eds.), *A companion to qualitative research* (pp. 203–208). Sage.

Instituto Distrital de las Artes. *Idartes.gov.co*, 2015. [Online]. Available: http://www.idartes.gov.co/index.php/programas/cumbre-mundial-arte-y-cultura-para-la-paz. Accessed 04 Sept 2015.

Kulig, J. C., Edge, D. S., Joyce, B., & Deer, R. (2008a). Understanding community resiliency in rural communities through multimethod research. *Journal of Rural and Community Development, 3*(3), 77–94.

Kulig, J. C., Edge, D., & Joyce, B. (2008b). Community resiliency as a measure of collective health status: Perspectives from rural communities. *The Canadian Journal of Nursing Research, 40*(4), 92–110.

Kulig, J., Edge, D., Townshend, I., Lightfoot, N., & Reimer, W. (2013). Community resiliency: Emerging theoretical insights. *Journal of Community Psychology, 41*(6), 758–775.

Lampis, A. (2013). Cities and climate change challenges: Institutions, policy style and adaptation capacity in Bogotá. *International Journal of Urban and Regional Research, 37*, 1879–1901.

LaRosa, M. J., & Mejia, G. R. (2012). *Colomiba: A concise contemporary history*. Rowman & Littlefield Publishers.

Longstaff, P. H., Armstrong, N. J., Perrin, K., May, W., & Hidek, M. A. (2010). Building resilient communities: A preliminary framework for assessment. *Homeland Security Affairs, 6*(3), 1–23.

Lüders, C. (2004). Field observation and ethnography. In U. Flick, E. von Kardoff, & I. Steinke (Eds.), *A companion to qualitative research* (pp. 222–230). Sage.

Luthar, S. S. (1991). Vulnerability and resilience: A study of high-risk adolescents vulnerability and resilience: High-risk adolescents. *Child Development, 62*(3), 600–616.

Luthar, S. S., & Zigler, E. (1991). Vulnerability and competence: A review of research on resilience in childhood. *The American Journal of Orthopsychiatry, 61*, 6–22.

Machlis, G. E., Force, J. E., & Burch, W. R. (2008). The human ecosystem Part I: The human ecosystem as an organizing concept in ecosystem management. *Society and Natural Resources: An International Journal, 10*(4), 347–367.

Masten, A. S., Best, K. M., & Garmezy, N. (1990). Resilience and development: Contributions from the study of children who overcome adversity. *Development and Psychopathology, 2*, 425–444.
Mcilveen, P. (2008). Autoethnography as a method for reflexive research and practice in vocational psychology. *Australian Journal of Career Development, 17*(2), 13–20.
Mcmillan, D. W., & Chavis, D. M. (1986). Sense of community: A definition and theory. *American Journal of Community Psychology, 14, no. January*, 6–23.
Medina, C., Morales, L., & Nuñez, J. (2008). Quality of life in urban neighborhoods in Colombia. *Borradores de Economía, 536*, 1–61.
Montezuma, R. (2005). The Transformation Of Bogota , Colombia , 1995–2000: Investing In Citizenship And Urban Mobility. *Global Urban Development, 1*(1), 1–10.
Morrow, V. (1999). Conceptualising social capital in relation to the well-being of children and young people: a critical review. *Sociol. Rev., 47*(4), 744–765.
Norris, F. H., Stevens, S. P., Pfefferbaum, B., Wyche, K. F., & Pfefferbaum, R. L. (2008). Community resilience as a metaphor, theory, set of capacities, and strategy for disaster readiness. *American Journal of Community Psychology, 41*(1–2), 127–150.
Olsson, C., Bond, L., Burns, J. M., Vella-Brodrick, D. A., & Sawyer, S. M. (2003). Adolescent resilience: A concept analysis. *Journal of Adolescence, 26*(1), 1–11.
Palacious, M. (2006). *Between legitimacy and violence: A history of Colombia, 1875–2002*. Duke University Press.
Pfefferbaum, B. J., Reissman, D. B., Pfefferbaum, R. L., Klomp, R. W., & Gurwitch, R. H. (2007). Building resilience to mass trauma events. In *Handbook of injury and violence prevention* (pp. 347–358). Springer.
Pickett, S. T., William, B., Dalton, S., Foresman, T., Grove, M., & Rowntree, R. (1997). A conceptual framework for the study of human ecosystems in urban areas. *Urban Ecosystem, 1*, 185–199.
Pretty, J. (2011). Interdisciplinary progress in approaches to address social-ecological and ecocultural systems. *Environmental Conservation, 38*(2), 127–139.
Programa de Gobierno | Portal Bogota| Bogota.gov.co, *Bogota.gov.co*, 2015. [Online]. Available: http://www.bogota.gov.co/gobierno/programa-de-gobierno. Accessed 27 Aug 2015.
Ripoll, J. L. (2009). Love in the time of Cholera: LGBT rights in Colombia. *SUR-International Journal of Human Rights, 11*, 72–89.
Rose, A. (1999). Defining and measuring economic resilience to earthquakes. *Disaster Prevention and Management: An International Journal, 13*(4), 307–314.
Rose, A. Z. (2005). 11. Analyzing terrorist threats to the economy: A computable general equilibrium approach. *The Economic Impacts of Terrorist Attacks*, 196–217.
Rutter, M. (1985). Resilience in the face of adversity. Protective factors and resistance to psychiatric disorder. *The British Journal of Psychiatry, 147*, 598–611.
Rutter, M. (1987). Psychosocial resilience and protective mechanisms. *The American Journal of Orthopsychiatry, 57*, 316–331.
Rutter, M. (2012). Resilience as a dynamic concept. *Development and Psychopathology, 24*(2), 335–344.
Safford, F., & Palacios, M. (2002). *Colombia: Fragmented land, divided society*. Oxford University Press.
Saskiewicz, P. (2005). *The revolutionary armed forces of Colombia – People's Army (FARC-EP) Marxist- Leninist insurgency or criminal enterprise?* Monterey.
Serrano-Amaya, J. S. (2011). Challenging or reshaping heteronormativity with public policies? A case study from Bogotá, Colombia. *IDS Working Papers, 1*(41).
Smith, G. (1999). Resilience concepts and findings: Implications for family therapy. *Journal of Family Therapy, 21*(2), 154–158.
Smith, B. W., Dalen, J., Wiggins, K., Tooley, E., Christopher, P., & Bernard, J. (2008). The brief resilience scale: Assessing the ability to bounce back. *International Journal of Behavioral Medicine, 15*, 194–200.

Sonn, C., & Fisher, A. (1988). Sense of community: Community resilient responses to oppression and change. *Journal of Community Psychology, 26*(5), 457–472.

Sotto, L. R. J., & Moreno, J. C. M. (2012). Construcción de subjetividades políticas en jóvenes LGBT de Chapinero–Bogotá. *Revista Iberoamericana de Psicología: Ciencia y Tecnología, 5*(1), 29–36.

UNOCD. (2013). *Global study on homicide 2013*.

Vale, L. J., & Campanella, T. J. (2005). *The resilient city: How modern cities recover from disaster*. Oxford University Press.

Wade, P. (2000). *Music, race, and nation: Música tropical in Colombia*. Univesity of Chicago Press.

Wade, P. (2006). Understanding 'Africa' and 'blackness' in Colombia: Music and the politics of culture. In K. Yelvington (Ed.), *Afro-Atlantic dialogues: Anthropology in the diaspora* (pp. 351–378). School of American Research Press.

Wall, S. (2008). An autoethnography on learning about autoethnography. *International Journal of Qualitative Methods, 5*(2), 146–160.

Waller, M. A. (2001). Resilience in ecosystemic context: Evolution of the concept. *The American Journal of Orthopsychiatry, 71*, 290–297.

Werner, E. E. (1995). Resilience in development. *Current Directions in Psychological Science, 4*(3), 81–85.

Wilkinson, C. (2011). Social-ecological resilience: Insights and issues for planning theory. *Planning Theory, 11*(2), 148–169.

Wolff, S. (1996). The concept of resilience. *The Australian and New Zealand Journal of Psychiatry, 29*, 565–574.

Zea, M. C., Aguilar-pardo, M., Betancourt, F., Reisen, C. A., Gonzales, F., Betancourt, F., & Reisen, C. A. (2014). Mixed methods research with internally displaced Colombian gay and bisexual men and transwomen. *Journal of Mixed Methods Research, 1*(10), 212.

Chapter 6
Resilience and Sustainability in Urban Socioecosystems: A Conceptual Reflection

Octavio Francisco González-Castillo

Abstract Chapter opens with a tight chronicle of 4.5 million years of evolution and humanization of the homo genus. In synchrony, the natural world in which it inhabits (real world or REALity) has also evolved, while a new and equally surprising world, of its creation, has emerged and has not stopped expanding its borders: its cultures or CONCEPTUALity. Since then, REALity and CONCEPTUALity have been intertwined, giving rise to authentic revolutions (agricultural, commercial, industrial, ICT, artificial intelligence, …) that, among other consequences, have enabled the exponential growth of their population, as well as the emergence—and since then, incessant growth—of the urban phenomenon. The rest of the chapter flows to answer the questions: what purposes has urban development served, under an anthropocentric approach? What new purposes should aspire to serve if development assumed a socioecocentric approach, oriented towards the resilience and sustainability of socioecosystems (wild, rural, and urban)? The author invites us to make the local and global crisis the starting point for the construction of a new development approach, one in which the final beneficiary for the development should be the socioecosystems, subject to a long-range planning and management (space, time, actors involved), in a participative orientation with an extended responsibility. To achieve this, the author concludes, it will be necessary to first change our cognitive and intervention approaches, as well as design novel methodologies and instruments to assess resilience and sustainability in socioecosystems.

Keywords Socioecocentrism · Urban socioecosystem · Resilience · Sustainability

O. F. González-Castillo (✉)
Metropolitan Autonomous University – Iztapalapa, Mexico City, Mexico
e-mail: ogc@xanum.uam.mx

O. F. González Castillo et al. (eds.), *Urban Resilience: Methodologies, Tools and Evaluation*, Resilient Cities, https://doi.org/10.1007/978-3-031-07586-5_6

6.1 Introduction

Since its origins, the human being has been faced with a daily battle for its survival. His genetic inheritance and the progressive development of the outermost layer or cortex of his brain—which endowed him with a degree of consciousness and reasoning with the ability to imagine, plan and learn—have allowed him to survive and adapt to a very wide range of circumstances and climate conditions, in very diverse, dispersed, and changing environments in practically all latitudes and longitudes of our planet, thus conforming its REALity or real world in which it inhabits and lives.

The gradual development of his ability to create symbols and communicate has allowed him to conceptualize, share and enrich its beliefs, knowledge, and skills,[1] as well as transmit these, generation by generation, creating and recreating various "cultures", that are shaping, in parallel to its daily REALity, an emergent dimension, of an atemporal nature, which we will call: CONCEPTUALity.

As an individual, the human being began to develop as an intelligent animal, yes, but also fragile and vulnerable to an environment that, while enables his life support, continually subject him to conditions and situations that put his existence at risk. His sexed nature and a herd instinct led him to form families, and then integrate them into increasingly numerous clans and communities. This community development became a real strength to face with advantage the continuous threats from the environment, likewise, it led to a progressive and increasingly efficient process of specialization-cooperation that would lead it to organize itself to intervene and transform its environment in an extensive and deep increasingly way.

One of the notable changes in the humanization process was that of establishing himself in a fixed territory, leaving behind the nomadic lifestyle, in which it was necessary for him to constantly mobilize to follow his hunting prey and to find new territories to collect the resources necessary for their subsistence. With a sedentary life, new needs, dynamics, and competition factors emerged (sense of property, accumulation, wealth, and power) that began to influence both the relationships between the different human groups, and between them and nature. The human being gradually became less dependent on his physical attributes (strength, speed, agility, etc.) to guarantee the expansion of the species.

His ability to imagine and build tools led him to transform his environment, and ultimately the planet, in a way that no other living being that we know of has done before. In just one instant of the evolution of the planet, the human being became a force of geological transformation of such magnitude that recently Paul Crutzen (2002) and Steffen et al. (2007) proposed that from the industrial revolution we are moving towards a new geological epoch, to which called *Anthropocene*, a geological epoch in which the human being has risen as a force with the ability to alter the destiny of the planet.

In the process of obtaining advantages to survive in the natural world, the development of the intelligence of the individual, as well as of the culture of the

[1] In the framework of Ethnoecology, Toledo-Manzur (1992, 2002) identify this set as the complex (k-c-p) kosmos (beliefs), corpus (knowledge) and praxis (skills).

communities, began to be a determining factor of accelerated expansion against the inexorable, but slower evolution process of the nature. With each one of the innovative technological revolutions triggered by human ingenuity (agricultural, manufacturing and trade, industrial, ICT), the way of life of the human species has been changing radically. Thus, it went from living in trees, caves, huts, and villages, to living in towns, cities, metropolitan areas and, currently, in true megalopolitan conglomerates.

From the incessant deployment, construction and cultural reconstruction and intervention on the environment, the human being has recreated his own built space, whose most complex expression is found in the cities. Currently the population exceeds 7850 million people, of which more than 4300 million (55%) live and develop their activities in urban areas (UN, 2020), surrounded by technological advances (technosphere), which in the framework of an economy based on capital and the market, they have promised to satisfy a set of needs, many of them never before imagined by man.

However, in recent decades we have become aware of the cost that in progressive levels of deterioration has been associated with "development", whose benefits, in fact, have reached in a very unequal way the set of individuals that in this,[2] the past, and even in future generations, we have shaped and will conform to the human species.

A sudden global awareness, which began with the book The Silent Spring (Rachel Carson, 1962), was coupled with a surprising accumulation of achievements and discoveries of the past 150 years[3] that ended up detonating, in the course of so only two generations, what some authors consider a global civilizational crisis. The urban phenomenon is no stranger to this crisis, which has deeply questioned the way in which we have approached, intervened, and reflected on the world.

[2] Ludevid-Anglada (1998, pp. 63 to 64, 113 and 139) offers the following contrasts:

(a) An average Latin American citizen has a life expectancy equivalent to that of a Western European in 1938; in the case of Asia, the equivalent year would be 1890 and for Africa, 1880. The contrast becomes more dramatic if we consider that the quality of life to which each of these countries can aspire is equivalent, respectively, to 50%, 20% and 25% of the quality of life enjoyed by a Western European citizen.
(b) It is estimated that the developed countries of the OECD (25% of the world's population) concentrate 85% of economic activity and 75% of energy consumption.
(c) Per capita energy consumption in industrialized countries is 14 times higher than in poor countries. Regarding its oil refining capacity, the factor is 24.

[3] Among others: evolution (Darwin, 1859); quantum model (Planck, 1900, 1901); relativity (Einstein, 1905, 1916); computer summit of Everest thermodynamics of irreversible systems/dissipative structures (Prigogine, 1947, 1967); General Systems Theory (Bertalanffy, 1968); first human being to step on the Moon (Armstrong, 1969); Arpanet → Internet (1969, 1984); environmental impact assessment resilience (Holling, 1973); steady-state economics (Daly, 1977); sustainability (WCED: Brundtland Report, 1987); transdisciplinarity (Nicolescu, 1987); World Wide Web (1990); cell phone (1993); complex adaptive systems (Holland, 1995); Dolly first cloned mammal (1997); Google (2000); Wifi (2003); Facebook (2004); Youtube (2005); e-phone (2007); Whats App (2009); Instagram (2010); Tik-Tok (2015); Covid-19 pandemic (2020).

6.2 Signs and Causes of the Problem

An orientation towards social reproduction,[4] as well as a development model that relies on the continuous growth of an economy, has ended up exceeding the carrying capacity of nature.

The problems associated with global warming, the destruction of the stratospheric ozone layer, deforestation and loss of biodiversity, acid rain, among others, are well documented in the scientific literature, so without pretending to be exhaustive, we will focus our attention on some other problems, associated with population growth, the dominant development model, as well as the trend towards urbanization.

In a development model whose aspiration and driving force is focuses on the incessant growth of economic wealth (measured by increasing GDP), it is not surprising that progressive and incessant formulas for growth in production and consumption are also promoted. If, in addition, the model has managed to permeate in the social imaginary (conceptuality) a tendency to "compete to have more things" as opposed to "collaborate to be fuller and happier", it is easy to understand why the dominant development model has derived in a model materialistic, self-centred and motivated by an ambition for financial/economic profit, which incessantly grows under the aspiration of increasing "having" and "power", which have risen as authentic competitive advantages in an economy based on the free market.

What better place to compete than in the huge markets represented by contemporary cities that in a few square kilometres gather huge populations of consumers, whose social imaginary has been "moulded" to promote an exacerbated consumption of an endless number of innovative products oriented to satisfy needs, many of these created?

Thus, hand in hand with the dominant development model, urbanization has been a growing reality in our societies.[5] The "city" has positioned itself as an aspirational model for many of the social actors and agents (civil society, governments, businessmen, associations, etc.) throughout the world. The urban way of life promises a set of opportunities, benefits, and comforts that in a world besieged by poverty[6] and lack of job opportunities make it true poles of attraction.

---

[4] By 1800 the world population was around a million inhabitants. With the Industrial Revolution, the exponential growth of the population gained a new impulse, requiring only 127 years to double and another 47 years to do it again. Currently, the rate of growth begins to stabilize, estimating that another 49 years will have to pass before doubling again and reaching 8000 million by 2023. It is estimated that between the years 2050 and 2100 humanity will reach a maximum population peak of between 10,000 and 12,000 million inhabitants.

[5] In 1800 2% of the world's population lived in cities. Currently it is estimated that the percentage has risen to 55% and according to UN estimates (2020), between 2050 and 2100 it could reach 68%.

[6] The World Bank (2020) reports: "… [about]… a tenth of the world's population lives on less than $ 1.9 a day … a quarter … below $ 3.2 … and more 40% … (almost 3.3 billion people) below $ 5.5". Additionally, it estimates that for the first time in more than 20 years extreme poverty will increase in 2020.

Undoubtedly, the benefits of the urban lifestyle have been real, although they have reached the population following a logic of pyramidal distribution in which people of a very wide base has received too little benefits and a very narrow cusp has received a lot, often in levels that only make sense under a lifestyle of exacerbated, superficial, and wasteful consumption.

Although they were ignored for some time, as a counterpart to the benefits generated by the dominant development model, high costs have been paid in pollution and deterioration – both environmental and social. Once again, the distribution of these costs tends to follow a pyramidal logic, but this time the very wide base has had to share the payment of the costs for benefits that are usually not within their reach, while a very narrow cusp evades, or pay very little for the benefits and privileges that usually enjoy in a wasteful way.

The iniquities described go beyond the urban limits since, although the benefits of the urban way of life are usually enjoyed within its limits, the costs (externalities) cross the peri-urban strip to reach, either in a directly or indirectly way, up to rural and wild places across the globe. Thus, in thermodynamic terms, cities function as negentropic systems: they grow and maintain their internal order at the expense of increasing the entropy of the surroundings and remote environments from which it is supplied and often, affects.

According to Toledo-Ocampo (2014, p. 179): "This urbanization process has important consequences for regional landscapes. The conversion of natural and agricultural landscapes to fully urbanized landscapes is the dominant trend today. These systems use or extract resources from adjacent regions and place increasing pressures on Earth's ecosystems. For this reason, creating resilient and sustainable metropolises today has become one of the critical tasks for scientists and urban planners".

The World Bank, in its Poverty and Shared Prosperity 2020 report, identifies three factors whose convergence is causing the current crisis:

1. <u>The COVID-19 pandemic:</u> It is estimated that the pandemic will cause a fall of between 5–8% of GDP per capita, which will add between 88 and 115 million more people to extreme poverty in 2020, a figure that could increase to 150 million by 2021. The poorest people are also more likely to work in occupations and sectors less compatible with social distancing…, which increases their risk of exposure to COVID 19, with the consequences that this entails for them health and income.
2. <u>Armed conflicts</u>: In 2020, the 37 economies formally classified as affected by situations associated with fragility and conflict are only home to around 10% of the world's population, despite the fact that more than 40% of the world's poor live there … Conflicts … lead to reduced access to education … increased numbers of deaths and injuries … stunted growth and deterioration in mental health.
3. <u>Climate change</u>: represents a serious threat in the medium term. Globally, an estimated 1.47 billion people live in areas at high risk of flooding, among them; approximately half (taking the international poverty line of USD 5.5/day) are poor. The World Bank Shock Waves report (Hallegatte et al., 2016) estimates:

"… by 2030 climate change could push another 100 million people into poverty. There is abundant evidence that people living in poverty or close to the poverty line are particularly vulnerable to shocks such as natural disasters … deteriorating asset quality … dependence on livelihoods based in agriculture and in ecosystems that are vulnerable to natural disasters … rising food prices during supply shocks … increased susceptibility to climate-related diseases, like diarrhea and malaria".

According to the World Bank report (2020):

> … the crisis particularly affects people who are already poor or vulnerable … those with less education … wealth… adequate social security … with insecure jobs … less skilled occupations … fewer mechanisms to deal with the crisis situation, such as savings.

The World Bank foresees that the crisis will prolong its effects in the future and that it will incorporate a new trend in which an increasing number of urban inhabitants fall into extreme poverty, a situation that traditionally has affected the majority of the population of the rural areas. That is why, in order to reverse this serious setback in development progress and poverty reduction, countries will need to prepare for a different post-COVID economy, allowing capital, labour, skills and innovation move towards new businesses and sectors.

Clayton and Radcliffe (1996, pp. 7 y 13) stated that:

> The size and complexity of the Earth system indicates that there may be a very large number of potential development pathways, of which a small subset may be sustainable for the human species. This, in turn, indicates that there may be a number of stable states, to varying degrees, as well as various ways to achieve them and that, therefore, there is more than one possible policy for a transition to a more sustainable way of life. … What is evident is that it is no longer possible, given the current extent of human activity, to avoid making decisions about how we want to interact with the planet.

6.3 Socioecocentrism

The REALity and CONCEPTUALity, in which the human being dwell, lives and create, are continually intertwined, and amalgamated to approximate what Le Roy (1927), Vladimir Vernadski (1945) y Teilhard de Chardin (1955) called *Noosphere,*[7] while some others preferred to associate it with concepts as: *Welstanchauung* (Dilthey, 1914), *Paradigm* (Kuhn, 1962), *Social Imaginary* (Castoriadis, 1975), among others. It is a sphere of intelligibility from which the human being comes into contact, interprets, intervenes, and reflects on himself and the world of which he is a part and depends for his subsistence and development (geosphere and biosphere).

In the human being, the link between the REAL and the CONCEPTUAL world is so close and strong that if one aspires to a future in which the REALITY changes,

[7] From gr. νόος nóos 'inteligencia1 and sphere (Royal Spanish Academy, 2020).

it will be necessary to begin by reviewing, questioning, deconstructing, and rebuilding that CONCEPTUALity that has brought it, for better and for worse, to the current situation. In the words of Russell Ackoff (1999), in a paraphrase of a phrase attributed to Albert Einstein, he warns us:

> It is not possible to create a plan by which a system of problems is controlled effectively, without first changing the thought patterns that produced those problems.

A second famous phrase, this time attributed to Ernst Friedrich Schumacher (quoted by Walsh & Vaughan, 1980, p. 52), warned us about the complexity of the search just begun:

> Nothing is more difficult than to become critically aware of the presuppositions of one's thoughts … All thought can be scrutinized directly, except for the thought through which we scrutinize.

Both phrases frame very well the challenge of the project RESILIENT CITIES: Re-thinking Urban Transformation. But what are the changes towards which the paradigms of resilience and sustainability are reorienting us?

After locating and characterizing various approaches to the movement towards sustainability and proposing *socioecocentrism* as a conciliatory approach, between anthropocentrism and ecocentrism, González-Castillo (2008) identifies the following premises and prescriptions to guide change:

- Reintegrate in our minds "human systems" and "environmental systems" and make the resulting *socioecosystems* the new pertinent system to guide development and sustainability.
- Expand the space-time horizons considered in cognition-planning-management processes.
- Expand, with principles of equity and cooperation, the participation to all stakeholders.
- Apply the precautionary, austerity and justice principles to human interventions.
- Expand the ambit and scope of transversal co-responsibility of all the actors and agents in any intervention act.
- Promote plural value systems that guide good living and harmonious coexistence.
- Promote a process of innovation-learning-teaching, as well as continuous improvement towards sustainability.

The first premise that proposes refocus on socioecosystems the new pertinent systems to guide development, resilience and sustainability are the core change propose in the socioecocentrism scope. This change in the object of study-intervention is not minor, since it presents us with a system with the following characteristics:

- These are complex adaptive systems
- That follow non-linear trajectories or dynamics
- Are subject to uncertainty, risk, and competition

- They integrate human (teleological systems) that follow development dynamics that must be kept in coevolution with the natural environment (teleonomic systems) of which they are part.

Assuming the seven complementary prescriptions of the socioecocentrism propose implies the need to accepting a framework that requires significant adjustments at the level of cognition, intervention, and reflection perspectives, as well as in the strategies, tactics, models, instruments, tools, and dynamics for collaborative work. In the next section will review some proposals oriented in this regard.

6.4 Adjusting the Collaborative Framework from Socioecocentrism[8]

Current reflection on sustainability has shown us that the crisis we are facing involves a set of interrelated problems within multiple problems, whose manifestations reach both the local, regional, and even global scales.

This crisis presents at least the following inherent complications: (a) the systems involved are multiple and very diverse, as well as their interaction very complex, (b) the dimensions and scales pertinent to the analysis are multiple, (c) the affected actors are multiple, as well as multiple are also their visions about the world and (d) the institutions involved are multiple.

The above complications condition that the approach and resolution of the problems consider, at least, an orientation of complex systems, a work beyond monodisciplinarity, a form of participation that involves the various actors and a transversal co-responsibility of public, private and social institutions involved.

Numerous authors have highlighted the need for a new way to investigate, it is worth mentioning for example: "science for a post-normal era" (Funtowicz & Ravetz, 1993) and "The mode 2 to investigate" (Gibbons et al., 1994). Although this relatively new mode of collaborative work is able to approximate knowledge of the world, it is usually rather motivated by the need (often urgent) to address complex problems that occur in the real world and that affect or threaten socioecosystems.

Below are some proposals aimed at reorienting the work framework towards socioecocentrism approach:

[8]This section is developed based on previous works presented in various forums by González-Castillo (2008, 2015a, b, 2016) and incorporates some updates when it is considered that they enrich or clarify the topic.

6.4.1 Adjustments in the Cognitive Perspectives

The paradigms of resilience and sustainability are still under development; however, they have already influenced cognitive processes in the following way:

- Questions a purely reductionist stance and propose to complement it with an emergent stance that identifies with systems thinking and complexity.
- Expand the relevant fields and dimensions in the studies, as well as their spatial and temporal horizons.
- Propose to complement the mono and multidisciplinary dynamics with those of inter and transdisciplinary orientation.
- Promotes involvement and continuous learning towards the sustainability of the community involved.

The fact that traditionally human systems and their development dynamics have been an almost exclusive field of study for social disciplines, while the study of natural systems and their evolutionary dynamics have been approached almost exclusively by natural disciplines. It results in a fragmented vision dominating our minds in which both systems are dissociated.

It is not possible today to deny the benefits of specialization, however, it would not be wise to deny its intrinsic limitations and stop facing the challenges and opportunities that the need to build a vision in which human and natural systems are reintegrated brings with it, in our mind. This will require complementing the disciplinary approach with more inclusive and contextualizing approaches, as well as building conceptual and communication bridges between both fields, starting from which to inaugurate shared cross-cutting paths and routes that are aimed at integrating paradigmatic frameworks, as well as a language shared to approximate the sustainability of development.

If today it turns out to be a valid aspiration to transform ourselves into a sustainable society, we will have to reintegrate human and natural systems into our minds. Given the complex nature of the relationships between the two systems, this will require complementing our reductionist perspective with an expansionist perspective that explains the phenomena in their totality and not in isolated parts, while considering the context and their interactions.

In the following quote, taken from Clayton and Radcliffe (1996: 9–11), it is well appreciated the fact that in order to tackle the challenge that the development sustainability paradigm faces, it will be necessary to integrate different disciplinary perspectives:

> Scientific research develops our understanding of the behaviour of natural processes that shape the environment, to track change, identify trends, and predict possible outcomes … Social and economic research is essential if we are to determine which economic instruments and politicians will obtain the desired results with the maximum of economy of means and with the minimum of adverse effects on other objectives… it is also necessary to estimate the probable social redistribution and other consequences of political decisions … Philosophical and psychological analysis it is essential to elucidate the mixture of rational and irrational assumptions that underpin the ethical decision-making process … Scientific, economic and philosophical perspectives … are each related to one aspect of the

> problem. If we want to understand this in itself, we need to have a way to integrate … [those perspectives].

The phenomena implicit in sustainability are too complex for deductive analysis and not random enough for statistical treatment. In this interface, it is the complex systems perspective that offers a more pertinent methodological approach, as well as a multidimensional matrix on which a large amount of information, coming from different disciplines and domains, can be integrated.

6.4.2 *Adjustments in Intervention Approaches*

Regarding the intervention processes, the following influences are identified related with the resilience and sustainability paradigms:

- Criticizes the radical positions of both preservationism and transformism, while ask for emerging and conciliatory approaches towards sustainability.
- Expands the scope of the groups involved, as well as the spatial and temporal horizons of the intervention projects.
- Guide towards a participatory design and execution with extended responsibility.
- Development is understood to be a participatory process that satisfies, with intra and inter-generational equity and justice, the needs of socioecosystems, as well as their agents and actors, always in harmony with the anthropoenvironmental systems of which they are part.

In this context, it is proposed that the study and management of socioecosystems must have cultural openness, broaden the participation of the actors involved and there must be a transversality in the institutions involved.

6.4.2.1 Intercultural Inclusion

Bell and Morse (1999, XII, 31, 151 and 155) affirm that the recognition that sustainability means different things to different people is a fundamental step to understand the nature of the problems we face:

> … the holistic and anthropocentric essence of sustainability continues to elude any attempt to analyse it objectively … sustainability is an organic and evolutionary construct of our minds and not an inorganic and static entity that can be physically tested.

Suárez-Molinar (2010: 182) calls for a glimpse of the "need to configure a 'pluralist universalism' that accounts for the heterogeneity of the world". Transdisciplinary approaches could contribute to the achievement of sustainability by consolidating transversal and vertical networks of actors capable of establishing fruitful dialogues aimed at the analysis, design, implementation, and evaluation of policies that allow progress towards sustainability, at the local, regional and global levels. In the same direction, León-Olivé (2010: 116) proposes to develop the notion of "social innovation networks", which, under a transdisciplinary orientation, integrate traditional and scientific knowledge to face social and environmental problems.

6.4.2.2 Expand Stakeholder Participation

The sustainability paradigm assumes the need to achieve, in conditions of justice and equity, the satisfaction of the legitimate needs of the group promoting the intervention, the internal stakeholders of the intervention system itself, and the external stakeholders (including representatives of natural systems and future generations), which could be affected during an intervention.

Of course, the participation of all affected actors multiplies the points of view in the debate, and therefore, the possible conflicts of interest. Although this complicates the already complex process of analysis and the establishment of agreements and consensus, there is no doubt that a timely incorporation of all those involved also brings the benefit of enriching the analysis, by incorporating legitimate points of view that, not being taken into account could reduce the sustainability of the intervention.

Under a socioecocentric approach, motivation would focus on serving, in a satisfactory and simultaneous way, the interests of all the agents, people or elements that are part and interact within the ecosystems and anthropoenvironmental systems involved.

6.4.3 Adjustments at the Level of Work Dynamics

The complex nature of socioecosystems conditions that the number and degree of complexity of the interactions that occur between their parts, as well as between the whole and its environment are high. In this case, the environment or suprasystem acquires a particular relevance since it will not only influence the behaviour of the system, but in a certain way, regulate it by positioning itself as the major system that is affected, or even more, must be served by the performance of the system itself.

Thus, the phenomena studied behave as if they were dependent on each other, and we could expect that the behaviour of the "whole" is more than the sum of the behaviour of its "parts".

Thus, in the face of socioecosystems, the motivation for collaborative work is usually curiosity and desire to know the performance of these complex systems in their environment, or to intervene in reality to increase the level of development and resilience of the socioecosystem.

6.4.3.1 Inter and Transdisciplinary Work Dynamics

Given the dynamic, complex and multidimensional nature of the relationships between natural and human systems, the understanding of sustainability requires collaborative work under inter and transdisciplinary dynamics between professionals and actors from very diverse fields.

The complex nature of socioecosystems conditions that the most appropriate approaches to address them, in their entirety, are those of inter and transdisciplinary nature. However, in the socioecocentric approach, its complementarity of these work dynamics is promoted, with those of a mono and multidisciplinary nature whose contributions tend to offer a starting point for the understanding of socioecosystems. Below is a summary of the dynamics of the different orientations for disciplinary work.

Monodisciplinary

- A skilled observer identifies and isolates a simple phenomenon for study.
- The nature of the phenomenon and its pertinent disciplinary field are determined.
- An experiment relevant to the study of the phenomenon is designed.
- A hypothesis is formulated to explain the phenomenon.
- An attempt is made to refute the hypothesis to delimit the NOT true until the truth is settled.
- A law is introduced that explains the phenomenon and is presented to the scientific community.

Polydisciplinary and multidisciplinary

- A problem situation is divided into apparently simpler situations.
- A team of specialists in the relevant disciplinary fields is integrated.
- Separate specialists analyses and solve simple situations.
- It continues in monodisciplinary cycles until all simple problem situations are addressed.
- It is assumed that adding the partial solutions will solve the original problem situation.
- At the end of the project the team of specialists is dispersed.

Interdisciplinary

- A complex problem situation is recognized in its global context.
- A team of specialists (EE) is integrated in pertinent disciplinary areas.
- The problem situation is addressed, in all its complexity, by each of the specialists.
- In-group session, each specialist gives their points of view and explains their results.
- In a hermeneutical-dialectical circle, EE integrates conclusions and agreements.
- At the end of the project the team of specialists is dispersed.

Transdisciplinary

- A core group is formed with representatives of the "owners" of the problematic situation (GND).
- Ideally, a core group of experts in basic areas (GNE) is also integrated.
- A common conceptual framework and shared language are amalgamated/enriched.
- A team of specialists (EE) in pertinent disciplinary areas and a representative group of actors (GA) are integrated.
- The complex problem situation in its global context is approached in a participatory way.
- In a hermeneutic-dialectical circle, GND, GNE, EE and GA meet to integrate conclusions and agreements.
- GD and GNE stay together, integrate the experience, and start a new learning cycle.

6.4.3.2 Transversal Co-responsibility of the Institutions

Under the socioecocentric approach, it will be necessary to redefine the rights and responsibilities assumed by the direct and indirect investors of the intervention, the internal and external human stakeholders of the intervention (including future generations), as well as the stakeholders that represent the interests of natural systems. A promoter under this approach would be committed to bringing the maximum benefits to the socioecosystems and anthropoenvironmental system, as well as promoting the acceptable, fair, and equitable distribution of these—as well as their costs—among all stakeholders and systems involved, at any scale, in the intervention.

González-Castillo (2008) characterizes the following intervention approaches based on the following description of their motivations:

- **Egocentrism**: development of a personal system for which use is made of both personal resources and inputs (goods and services) from human and natural environments.
- **Anthropocentrism**: development of a human system for which both personal resources and goods and services of human and natural environments are used.
- **Preservationism**: preservation of a natural system.
- **Eclectic transformism**: development of a personal or human system, controlling the impact on ecosystems or socioecosystems of the environment.
- **Eclectic conservationism**: conservation of a natural system, controlling the impact on sociosystem or socioecosystems of the environment.
- **Local socioecocentrism**: development of a local socioecosystem, maintaining the resilience and adaptability of the socioecosystems with which it maintains direct interaction in its environment.
- **Global socioecocentrism**: development of a local socioecosystem, maintaining the resilience and adaptive capacity of the planet's anthropoenvironmental systems.

Thus, sustainability confronts "development models" and their institutions with the need to assume co-responsibility to move from an exacerbated EGOcentric model to a SOCIOECOcentric model.

All of the above indicates that we are facing a true cultural revolution, a reengineering of the civilizational model that should guide us in the transition towards the sustainability of socioecosystems, under approaches to internalize an anthropoenvironmental responsibility. In this regard, Clayton and Radcliffe (1996: 6) warn:

> … the number, complexity, and interrelationships among the institutions involved indicate that a strategy consisting of relatively disconnected adjustments from each other and from economic and social policies and means … is less likely to be successful than a systematic attempt to build a socio-economic system that fits and interacts appropriately with the planet's environmental systems … Such an approach should include the evaluation and reassessment of most social and economic objectives. In fact, the authors continue, many of the existing organizations, as well as political and economic concepts and structures, today are probably inappropriate … and some organizational and political transformation will be necessary. This in turn is unlikely to happen if it is not accompanied by a parallel evolution of the cultural and psychological concepts on which political and social structures are ultimately based.

6.4.4 Adjustments at the Level of Models and Instruments

In order to approximate the complex nature of socioecosystems, several inter and transdisciplinary fields have been emerging, among others the following can be highlighted:

- Ecological design.
- Ecological economics.
- Environmental economics.
- Environment education.
- Environmental ethics.
- Environmental law.
- Ethnoecology.
- Human ecology.
- Landscape ecology.
- Political ecology.

Each of these interdisciplinary fields has developed its own work methodologies (see for example Toledo-Ocampo, 1998). The description and application of several of them can be consulted by the reader in the cases presented in the various chapters of this book. A similar situation occurs with respect to the multiple models, instruments and work tools developed in each of these fields for the study and management of socioecosystems, just to mention a few:

- Big data analysis.
- Discourse analysis and hermeneutics.
- Cartography and Geographic Information Systems (GIS).
- Cartography and Conceptual Information Systems (CIS). (González-Castillo, 2008)
- Circular economy.
- Heuristic methods.
- Interviews and questionnaires.
- Life cycle analysis.
- Maps of actors.
- Modelling and simulation of dynamic systems.
- Policy analysis.
- Resilience and sustainability assessment.
- Scenarios and game theory.
- Study of cases.
- Uncertainty and sensitivity analysis.

The reader is again invited to review the different chapters of this book, in which he will find numerous practical cases of the use of these instruments.

6.5 Epilogue

Socioecocentrism emerges as the proposal of a new development approach towards sustainability, in which, to reconcile the extreme positions of anthropocentrism and ecocentrism, socioecosystems are conceptualized and proposed as the centre or focus point of the intervention intentions. Under this new approach, a style of human development is promoted that harmonizes with the processes of coevolution typical of local socioecosystems, as well as of the planet's anthropoenvironmental systems. Thus, it will be the legitimate interests of the socioecosystem that will serve as a target and guide to define and approximate the notions and instrumental definitions for both, the "development" and the "sustainability".

The minimum scale of analysis to evaluate the impact on resilience and sustainability implicit in the performance of a productive system is that of the socioecosystem, of which the former is a part. If the migration of impacts (positive or negative) to other neighbouring socioecosystems is foreseen, the analysis should also be carried out at the level of the affected regional or planetary anthropoenvironmental systems.

In all human intervention on the socioecosystem, the quality of life, resilience, and adaptability of both the socioecosystem (local) and the anthropoenvironments (regional and global) of which the former is a part should be valued, maintained and if possible, incremented.

The quality of life, resilience and global capacity for adaptation in urban socioecosystems emerge as complex properties and can be increased through actions aimed at reducing the vulnerability of the socioecosystem and increasing its resilience in an articulated, flexible and synergistic manner in three areas: a) its infrastructure (tangible and intangible), b) its social fabric (individuals, legal entities) and their cultures and c) the natural support base on which its well-being depends and with which it co-evolves.

In all cases, it will be necessary to take into account and, if necessary, extend the scope of actions towards those other urban, rural or wild socioecosystems with which the urban socioecosystem in question maintains interactions, whether the latter occur directly or indirectly.

Under the previous premises, a socioecocentric approach shares and assumes to a greater or lesser extent the following set, which does not pretend to be exhaustive, of convictions, values, and principles:

1. CONVICTIONS.

   1.1 <u>Socioecosystems.</u>

      1.1.1 The development of human systems cannot be separated from the evolution of natural systems. Both systems co-evolve in exchanging flows of matter, energy, and information.

      1.1.2 Biological and cultural diversity, in addition to have value itself, is a key factor that contributes to the stability of life support systems on the planet.

   1.2 <u>Pollution and environmental deterioration.</u>

      1.2.1 The phenomena of contamination, extinction of species, poverty and, in general, socioenvironmental deterioration represents anomalies within the current development model.

      1.2.2 Technical solutions are necessary, but not sufficient to achieve sustainability.

   1.3 <u>Movement toward sustainability.</u>

      1.3.1 The paradigms of environmental care (years 70–80) and sustainability (years 80–90) emerged at the end of the twentieth century as two complementary and progressive responses to face the environmental crisis of the planet.

      1.3.2 The paradigm of sustainability emerges as the proposal of a new scenario for world development, in which the idea of development with stability must be satisfied.

      1.3.3 The paradigm of sustainability represents a profound transformation that questions the way in which human beings, in the search for the satisfaction of their needs, have observed, thought about and intervened on the planet.

1.3.4 Sustainability can be considered as an unattainable ideal, but it is convenient to try to approximate it; it is a dynamic mental construct (man-designed abstract system) that must be continually discussed, agreed upon and revised locally.
1.3.5 It is at the level of "approaches" that the relevant debate on sustainability must take place. That is, to understand that what sustainability puts at stakes are all those unquestioned assumptions that underpin the set of paradigms that make up (or will make up) the different contending worldviews.

1.4 <u>Who will be considered as legitimate actors or agents in an intervention?</u>

1.4.1 The productive systems must be oriented towards satisfying, with inter- and intragenerational equity and justice, the needs of all legitimate stakeholders within the socioecosystems and anthropoenvironmental systems involved.
1.4.2 The largest possible number of legitimate stakeholders should be considered and involved: achieving sustainability requires the participation and commitment of all social sectors.
1.4.3 A process of analysis and participatory decision-making is promoted.
1.4.4 A commitment must be established with future generations so as not to restrict their right to access socioenvironmental resources, at least equivalent to that of the present generation.

2. VALUES.

2.1 Natural and human systems have value in themselves and are not just exchangeable commodities on the market.
2.2 Equity and inter and intra-generational justice are promoted.
2.3 In any intervention, the search for effectiveness (achieving the objective) and efficiency (with the least possible expenditure of resources) must have ethical support that confers legality and legitimacy.
2.4 Scientific and technological progress must be subject to moral and ethical principles to guarantee both the research procedures used and the responsible use of the basic and applied knowledge that emanates from it.
2.5 A culture of plural and inclusive values must be promoted and prevailed, as well as a system of extended responsibility that reaches all the actors and agents involved in any act of intervention.

3. PRINCIPLES.

3.1 <u>Structural principles.</u>

3.1.1 Maintain the quality of life, resilience and adaptability of both the local socioecosystem under study-intervention, and the anthropoenvironments (regional, global) of which the former is part.
3.1.2 Maintain the biological and cultural diversity of socioecosystems.

3.1.3 Protect the integrity of the structures and processes on which the life support of the planet depends.
3.1.4 Avoid contamination: reduce, recover, reuse, and recycle.
3.1.5 Think globally and act locally: faced with the challenge of sustainability, each socioenvironmental community has the right to develop its own local solutions, but it must also assume the responsibility to contribute to and preserve the global sustainability of the planet.
3.1.6 Make our production and consumption patterns more austere.
3.1.7 Addressing situations of poverty and malnutrition, as well as restoring damaged ecosystems to guarantee the health and integrity of the socioecosystem, acquires priority.
3.1.8 Articulate international cooperation strategies to solve global problems.
3.1.9 Establish population policies according to the carry capacity of the socioecosystems to support development with quality of life.
3.1.10 Establish equitable, fair, and open trading systems.
3.1.11 Value production for self-consumption.
3.1.12 Establish a price system that incorporates the value of environmental goods and services.
3.1.13 Create mechanisms of justice and equity in the distribution of development benefits and costs.
3.1.14 Developing a culture aimed at valuing and caring for socioecosystems.
3.1.15 Engage citizen participation.
3.1.16 Apply the precautionary principle.

3.2 Operative principles.

3.2.1 Consume renewable resources at rates lower than nature's ability to regenerate them.
3.2.2 Consume non-renewable resources at rates lower than our capacity to generate substitutes.
3.2.3 Do not dump non-biodegradable waste into nature.
3.2.4 Prefer the use of solar and wind energy, as well as the consumption of clean fuels.
3.2.5 Do not dump biodegradable waste into nature at a speed greater than its carrying capacity.
3.2.6 Design efficient, cyclical and environmentally friendly production processes.
3.2.7 Increase the efficiency of energy use.
3.2.8 Reduce, reuse, and recycle man-made waste.

References

Ackoff, R. (1999). *Recreation of corporations: An organizational design for the 21st century*. Oxford University Press.

Bell, S., & Morse, S. (1999). *Sustainability indicators: Measuring the immeasurable? Ed.* EarthScan Publications Ltd.

Bertalanffy, L. (1968). *General system theory: Foundations, development, Applications* (p. 1976). George Braziller.

Carson, R. (1962). *Silent spring*. Penguin.

Castoriadis, C. (1975). *The imaginary institution of society*. MIT Press.

Clayton, A. M. H., & Radcliffe, N. J. (1996). *Sustainability: A system approach*. Ed. Earthscan.

Crutzen, P. J. (2002). *Geology of mankind* (Nature) (Vol. 417, p. e23). Macmillan Publication.

Daly, H. E. (1977). *Steady-state economy*. Freeman.

Darwin, C. (1859). *On the origin of species*. John Murray.

De Chardin P. T. [1955]. *The human phenomenon*. Harper & Brothers (1959).

Dilthey, W. (1914). *Einleitung in die Geisteswissenschaften. Versuch einer Grundlegung für das Studium der Gesellschaft und der Geschichte*. Vandenhoeck & Ruprecht.

Einstein, A. (1905). Zur Elektrodynamik bewegter Körper. *Annalen der Physik (Berna), 17*, 891–921.

Einstein, A. (1916). The foundation of the generalized theory of relativity. *Annalen der Physik, 354*(7), 769–822.

Funtowicz, S., & Ravetz, J. (1993). Science for the post-normal age. *Futures, 25*(7), 739–755.

Gibbons, M., Limoges, C., Nowotny, H., Schwartzman, S., Scott, P., & Trow, M. (1994). *The new production of knowledge: The dynamics of science and research in contemporary societies*. SAGE.

González-Castillo, O. (2008). Diseño de un instrumento de planeación Para incorporar, en la evaluación de sistemas, enfoques hacia la sustentabilidad. Thes is to obtain the Degree of Doctor, Faculty of Engineering, UNAM.

González-Castillo, O. (2015a). Movimientso global y organizacional hacia la sustentabilidad: Escuelas, enfoques y paradigmas. In *Course notes. Postgraduate in sciences and arts for design*. Xochimilco Unit. Metropolitan Autonomous University.

González-Castillo, O. (2015b). Socioecocentrismos: Propuesta de un nuevo enfoque hacia la sustentabilidad. In *Course notes. Postgraduate in sciences and arts for design*. Xochimilco Unit. Metropolitan Autonomous University.

González-Castillo O. (2016). *Sustentabilidad y la Universidad del siglo XXI: Aproximaciones al Conocimiento y al Trabajo Colaborativo* In. Laguna-Sánchez G.A. (Coordinator) *Complejidad y Sistemas Complejos: un acercamiento multidimensional*. Open-access book. Editorial Copit Arxives UNAM.

Hallegatte, S., Bangalre, M., Bonzanigo, L., Fay, M., Kane, T., Narloch, U., Rozenberg, J., Treguer, D., & Vogt-Schilb, A. (2016). *Shock waves: Managing the impacts of climate change on poverty* (Climate change and development series). World Bank Group.

Holland, J. H. (1995). *Hidden order. How adaptation builds complexity*. Perseus Books Group.

Holling, C. S. (1973). Resilience and stability of ecological systems. *Annual Review of Ecology and Systematics, 4*, 1–23.

Kuhn T. (1962). *The Structure of Scientific Revolutions*. Chicago, IL: University of Chicago Press.

Le Roy, E. (1927). *L'Exigence idéaliste et le fait l'évolution*. Boivin.

León-Olivé, M. (2010). Conocimientos tradicionales e innovación: desafíos transdiscipliarios. In Peláez, A., & Rodolfo, S. (Coords.), *Observaciones filosóficas en torno a la transdisciplinariedad*. Num. 16 New Science Collection. Ed. Anthropos. Rubí (Barcelona, Spain) and Metropolitan Autonomous University – Cuajimalpa.

Ludevid-Anglada, M. (1998). El cambio global en el medio ambiente: Introducción a su causas *humanas*. Ed. Alfaomega-Marcombo.

Nicolescu, B. (1987). *La Transdisciplinarité*. Editions du Rocher.

Planck, M. (1900). Zur theorie des gesetzes der energieverteilung im normal spectrum. *Verhandlungen der Deutschen Physikalischen Gesellschaft, 2*, 237–245.

Planck, M. (1901). Über des gesetzes der energieverteilung in normal spectrum. *Annalen der Physik, 309*(3), 564–566.

Prigogine, I. (1947). Etude thermodynamique des phénomines irreversibles. Bruxelles, published by Desoer, Liege.

Prigogine, I. (1967). *Structure, dissipation and life*. Communication at the first international conference: Physique Théorique et Biologie (ed. Marois) Amsterdam.

Royal Spanish Academy (RAE in spanish). (2020). Dictionary. https://dle.rae.es/noosfera. Accessed 17 Jan 2020.

Steffen, W., Crutzen, P. J., & McNeill, J. R. (2007). The Anthropocene: Are humans now overwhelming the great forces of nature? *AMBIO A Journal of the Human Environment, 36*(8), 614–621.

Suárez-Molinar, R. R. (2010). *Sobre los usos sociales del concepto transdisciplina*. En Peláez A. y R. Suárez (Coords.), *Observaciones filosóficas en torno a la transdisciplinariedad*. Num. 16 New Science Collection. Anthropos Editorial. Rubí (Barcelona, Spain) and Metropolitan Autonomous University – Cuajimalpa..

Toledo-Manzur, V. M. (1992). What is ethnoecology? Origins, scope and implications of a rising discipline. *Etnoecológica, 1*(1), 5–21.

Toledo-Manzur V. M., (2002). Ethnoecology: A conceptual framework for the study indigenous knowledge of nature. En Stepp, J.R., F.S. Wyndham, & R.K. Zarger, (eds.), Ethnobiology and biocultural diversity (pp. 511–522). : The International Society of Ethnobiology.

Toledo-Ocampo, A. (1998). *Economía de la Biodiversidad*. Series: Textos Básicos para la Formación Ambiental No 2; United Nations Environmental Program (UNEP), Regional Office for Latin America and the Caribbean. United Nations (UN).

Toledo-Ocampo, A. (2014). *Planning of complex socio-ecological systems. Science and society collection* (Science diffusion series). Autonomous University of Mexico City and Jorge L. Tamayo Engineering Center for Geography and Geomatics Research (GEO center).

United Nations (UN). (2020). Urban areas are home to 55% of the world's population, and that figure is expected to increase to 68% by 2050. Electronic publication October 20, 2020. Consulted at https://www.onu.org.mx/las-zonas-urbanas-already-home-to-55-of-the-world's-population-and-that-number-is-expected-to-rise-to-68-by-2050

Vernadski, G. (1945). *Biosphere y Noosphere*. English translation from the Russian original published in 1944. A few words about the noosphere. USSR. American Scientist.

Walsh, R., & Vaughan, F. (Eds.). (1980). *Beyond ego: Transpersonal dimension of psichology*. J. P. Tarcher. USA.

World Bank. (2020). *Poverty and shared prosperity 2020: Reversals of fortune* (Biennial report). World Bank.

World Commission on Environment and Development (WCED – UN). (1987). *Our common future* (Brundtland report). United Nations.

Chapter 7
System Approach to Resilience-Based Design: Political Decisions and Steps Towards Antifragility

Konstantinos Gkoumas, Francesco Petrini, and Franco Bontempi

Abstract In recent years, natural disasters are recognized to be the cause of considerable human and socioeconomic losses, particularly in modern, infrastructure-dependent societies. For example, the 2011 earthquake and tsunami in Japan have been one of the most devastating disasters of the past decades. Likewise, the Katrina hurricane was in the US east coast in 2005. On the other hand, climate change is considered a major issue nowadays and its consequences have been considered only recently in risk assessment. In this context, the concepts of "resilience of urban areas" and "resilient community", have gathered the attention of researchers. On top of that, more recently, antifragile design came as an evolution of design for resilience (intended as the capacity to recover), or for structural robustness (a main dimension of resilience, intended as the ability of a structure to withstand events without being damaged to an extent disproportionate to the original cause). This study focuses on a modern approach in disaster resilience – including issues for the risk assessment in antifragile design – providing insight and a framework on important modelling aspects, in particular:

(i) System representation, including hazard modelling and the time horizon of events;
(ii) Decision representation (with specific reference to political decisions);
(iii) Data collection, necessary for data-driven modelling.

The intention is to provide an umbrella framework and set of advises of correct practice that can help policy advisors and experts design for resilience in urban areas.

K. Gkoumas (✉)
Sapienza University of Rome, Rome, Italy

European Commission, Joint Research Centre, Ispra, Italy
e-mail: konstantinos.gkoumas@ec.europa.eu

F. Petrini · F. Bontempi
Sapienza University of Rome, Rome, Italy
e-mail: francesco.petrini@uniroma1.it; franco.bontempi@uniroma1.it

O. F. González Castillo et al. (eds.), *Urban Resilience: Methodologies, Tools and Evaluation*, Resilient Cities, https://doi.org/10.1007/978-3-031-07586-5_7

Keywords Resilience-based design · System quality · Political decisions · Structural robustness · Hazard representation

7.1 Introduction

In his reference book *Antifragile: Things That Gain from Disorder*, Nicholas Nassim Taleb explores the attribute of antifragility for systems (economic, social, natural etc.), as a step forward from robustness and resilience (Taleb, 2012). While fragile systems suffer or break from randomness and volatility, and resilient systems have the characteristic to stay the same, anti-fragile systems gain and grow stronger from variability and stress (up to a certain point). Taleb argues that instead of seeking to eliminate variability (something that can be perceived as a "loser's game", since variability and randomness are the rule and not the exception in everyday life), it is better to live and deal with it, and try to gain using different tactics.

In a different context, Italo Calvino, in his touchstone fabulist novel *The invisible cities* (Calvino, 1974), in a dystopian context (cities mostly represent dystopian urban environments), finds reminiscence and a sense of hope in fictional conversations between a young Marco Polo and ageing emperor Kublai Chan. Cities in his book represent complex historical examples and imaginary possibilities, characterized by their infinite complexity, their intensive urban landscape, and their strong interactions between them and their inhabitants. While some of them are utopian models of success, the majority of them are left to their destiny, being responsive to their purpose and to the acts of their inhabitants. What emerges is the idea that some cities are "invisible", ever changing, with details ready to be discovered (or left behind): in this sense, people continue to live in the cities, albeit the deficiencies or crisis situations. Calvino's envisions match Castoriadis' thoughts on society as a self-organizing entity. At the same time, moving beyond a concept of the city that merely on physical objects, would lead to an enhanced resilience as a cultural praxis (Castoriadis, 1987).

The above examples provide a starting point for discussions on the future of resilience assessment for urban developments, considering the complexity, the continuously changing aspects and the multiplicity of situations that can occur. In fact, cities and urban settlements tend to become more populated and complex, with the introduction of structures and infrastructures hardly imagined years ago (e.g. mega skyscrapers, extreme long span bridges).

Focusing on Fig. 7.1, Resilience can be perceived as an extension of robustness (a term common in the structural engineering community) and as a complement of antifragility. In the same figure, a timeline of significant events that led to the development of the concepts is also shown. This timeline is not exhaustive; however, it provides the reader with some turning points in the perception of these concepts.

In an urban settlement consisting of human-created structures, robustness refers to a single structure (or a series of structures, especially when considering the case of progressive or disproportionate collapse). Robustness provisions in modern

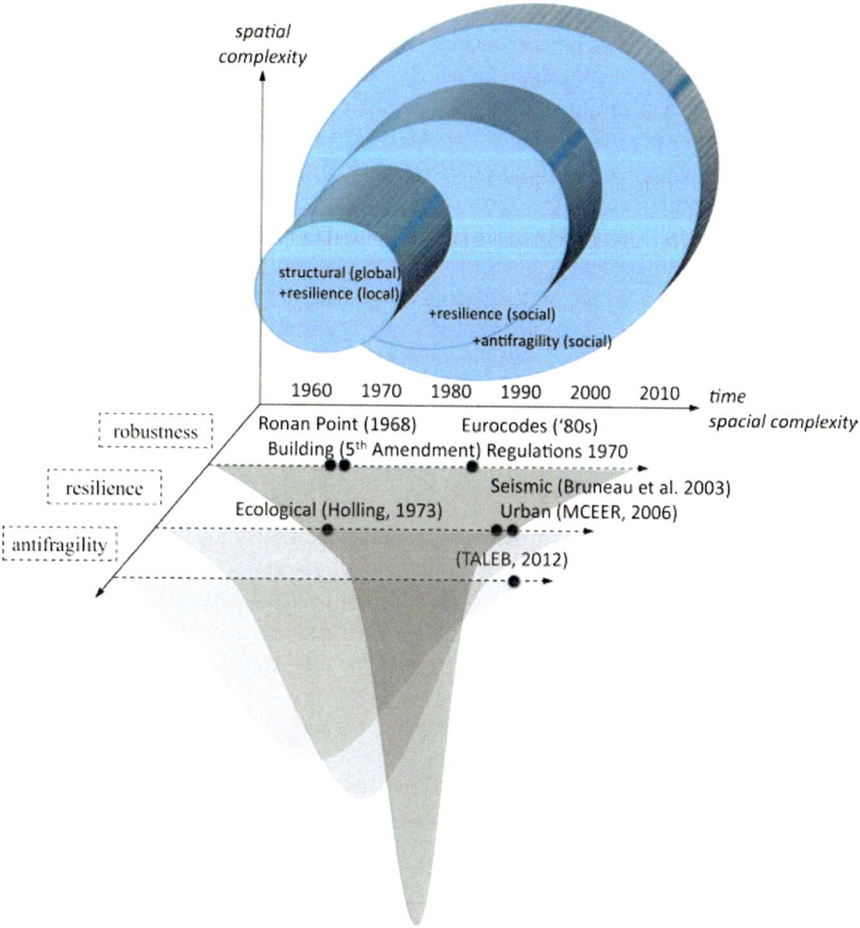

Fig. 7.1 Robustness, resilience and antifragility

structural design came into the spotlight after the Ronan Point collapse (1968) and many countries have adopted structural integrity or "robustness" provisions that may be directly traced to the Ronan Point collapse (Pearson & Delatte, 2005). Antifragility and resilience on the other side, refer to a complex of structures in the widest sense, including issues that go well beyond structural design. Ecological resilience (introduced in a consistent manner in Holling, 1973) is also nowadays trending for climate change provisions. Urban Resilience (including seismic), were introduced in Bruneau et al. (2003) and MCEER (2006). On top of that, antifragility (Taleb, 2012), is a novel concept, introduced only recently, that provides a new insight in risk assessment methods in different engineering and life science fields.

It is important to link the above three concepts in the field of study of interest (in this case, principally civil and structural engineering), and highlight their limits and dependencies.

- Structural robustness relates to a single structure or complex of structures. In this sense, it stays well above the concept of structural resistance (referred to a part of the structure or a structural element), but it is limited to a (even though large) structural system. In this sense, we can talk about the structural robustness of a high-rise building, a bridge, a hospital etc.
- Resilience is a much more complex term, that relates to broader systems (also socio-economic) well beyond structural measures. In civil engineering and architecture, we talk about the resilience of urban developments, even cities, or, depending on the threat consequences, even at a state level.
- Antifragility is very recent term that can find application in the risk assessment of complex systems. Even though initial applications are at the material level, the challenge is to adopt the concept for complex urban developments. In fact, it would be interesting to see cities, urban developments and structures not only recover, but also grow stronger after adverse events (floods, earthquakes, terrorist attacks).

Furthermore, robustness refers to the capacity to withstand (eventually with losses). Resilience to the capacity to recover (usually to the original state). Antifragility extends to bouncing back and improving the previous state.

This study, after some initial considerations, aims to address the consideration of different hazards (including climate change) to the political decisions that influence the society wellbeing both in the short and in the long term.

7.2 Resilience-Based Design

The concept of resilience is present since the 70's in fields of study such as psychology and ecology. The Merriam Webster dictionary defines resilience as "the ability to become strong, healthy, or successful again after something bad happens" (for individuals) or "the ability of something to return to its original shape after it has been pulled, stretched, pressed, bent, etc." (for objects or things).

The American Psychological Association (n.d.) defines resilience as "the process and outcome of successfully adapting to difficult or challenging life experiences, especially through mental, emotional, and behavioural flexibility and adjustment to external and internal demands". Even though psychological resilience is related to optimistic attitude and positive emotionality, it relies, among else, on adapting and on perspective, and some of the methods for building resilience can be applied also in other fields.

In ecological systems, resilience is defined as the capacity of an ecosystem to respond to a perturbation or disturbance by resisting damage and recovering quickly (see for example, Holling, 1973; Gunderson, 2000; Gunderson & Holling, 2002).

Even though resilience is domain-dependent (that is, it relates to the specific context), there are similarities in different fields and contexts.

Resilience has found application in the last years in other fields (e.g. electronic and computer systems). Maruyama et al. (2014), provide a taxonomy for general resilience, focusing on four orthogonal dimensions:

(i) Type of shock or perturbation
(ii) Target system
(iii) Phase of concern
(iv) Type of recovery

The same authors highlight strategies for achieving resilience (through redundancy, diversity or adaptability).

Davoudi et al. (2013) develop a conceptual framework by drawing on three broad and distinctive perspectives on resilience (engineering, ecological and evolutionary). Among their conclusions, they highlight the potential transformative opportunities that emerge from change.

In the civil and architectural engineering field, resilience is present through the notions of "resilience of urban areas" and "resilient community", as introduced by the Multidisciplinary Centre for Earthquake Engineering Research – MCEER (2006). The approach has the potential to provide a considerable contribution in lowering the impact of disasters, and is implemented through the Resilience-Based Design (RBD) for large urban infrastructures (buildings, transportation facilities, utility elements etc.), conceived as a design approach aiming at reducing as much as possible the consequences of natural disasters and other critical unexpected events on the society. Resilience based design can be considered as an extension of Performance based design in civil engineering, in which a structure needs to meet certain measurable or predictable performance requirements for efficiency, serviceability or strength without a specific prescribed method by which to attain those requirements. The goal of RBD is to make individual structures and communities as "Resilient" as possible, developing technologies metrics and actions that allows each structure and/or community to regain its function as promptly as possible (Bruneau et al., 2003; Renschler et al., 2010; Cimellaro et al., 2010; Cimellaro, 2016).

Several work focused on integrating or extending the above. Franchin and Cavalieri (2015), focusing on earthquakes, extend a previously developed civil infrastructure simulation framework to the evaluation of resilience, and introduce a new infrastructure network-based resilience metric. Their model builds on findings from Asprone et al., 2013, who introduce a metric with reference to the ability of a whole system to recover the full functional level, in terms of housing reestablishment, existing prior to the event even if in a new, different state.

Considering the above issues, it is safe to say that he concept of resilience implies multidisciplinary aspects (Bosher, 2008), and requires collaboration between different experts (e.g. urban designers, ecologists, engineers, architects, social scientists). Its definition is not univocal: a straightforward one is the one given by Lindell (2010), where a resilient community is defined as the one having the ability to absorb disaster impacts and rapidly return to normal activities. Furthermore, it has been recognized that some infrastructures are critical for resilience in the sense that they mostly contribute to the response to disasters. In a Resilience-based design, focus should be given to such infrastructures, but no criteria have been established yet for determining the role of different types of infrastructures in the achievement of a resilient response of an urban area to critical events.

Nowadays, urban resilience focuses on three distinct threats (Coaffee, 2008):

- Climate change
- Natural disasters
- Terrorism

Regarding threats from natural hazards in particular, resilience-based design focuses on possible outcomes from threats such as heat waves, droughts and flooding, earthquakes, tsunamis, solar flares, etc. The complexity arises from the possible occurrence of multiple hazards (eventually as a consequence of each other).

7.2.1 System Representation

An important aspect of resilience-based design is the representation of complex systems. In fact, theories of ecological and social-ecological systems embrace resilience as dynamic and operating in non-equilibrium (for a body of scholarship on resilience and complexity in such systems, refer to Meerow & Newell, 2015).

A physical infrastructural and human system in an urban area or community can be represented as an evolving hierarchical system (Salthe, 1985). In particular, the co-evolution of the system and the surrounding environment has to be considered.

A specification of an infrastructural and human system can be referred to as Complex, Large-scale, Interconnected, Open, Socio-technical (CLIOS) system, including elements from the built environment and the social-political domains (Mostashari & Sussman, 2009).

The system resilience is provided by two main characteristics: the *survivability* and the *elasticity*.

The concept of survivability is directly related to an appropriate measure of the system quality. The quality can be, for example, defined as the ability to maintain a certain level of safety and functional performance. The following examples of quality requirements can be given for specific infrastructural elements of an urban system (Ortenzi et al., 2013).

1. An aqueduct serving an urban area is required to maintain a certain level of usable water flow intended as functional quality.
2. A network of hospitals is required to maintain a certain capacity in receiving patients and an acceptable connection level between the single operating hospitals.
3. For an ordinary building, the safety of the occupants is the appropriate quality requirement, also maintaining the serviceability for the building evacuation paths.

The *complexity* of the system, the interaction between the subsystems, and the level of coupling between them are well recognized and classified as described in Fig. 7.2 (Perrow, 1999).

At this point, the system can be represented by an appropriate influence map. The components of the system are basically elements and connections between them (Bontempi et al., 2008). Elements are the main premises (structures and infrastructures) of the system (physical component), and the main human actors (social component) grouped in typologies (Cavalieri et al., 2012). The connections between elements are the physical and technical links between premises and social-hierarchical influences between human groups. The influence maps allow representing the connections between the elements and the type of mutual influences between the elements (how a certain loss of quality of an element can influence the quality of another element connected with the first one, or how a certain action from a group of human actors can condition the behaviour of another group).

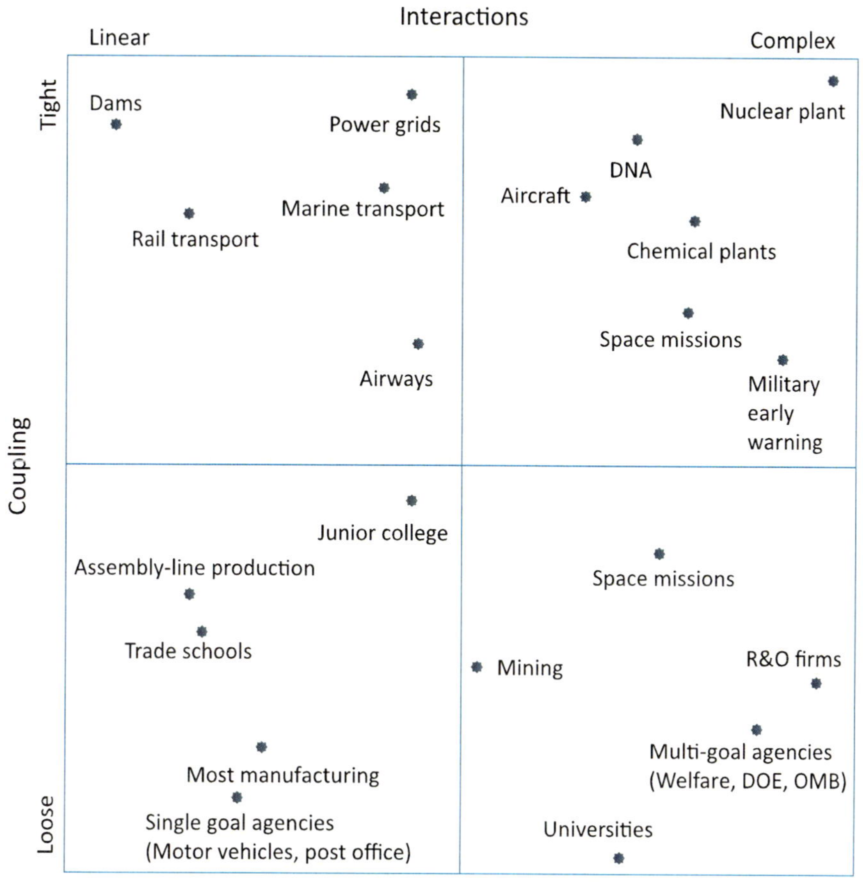

Fig. 7.2 Interaction between subsystems. (Source: Adapted from Perrow, 1999)

The physical component of the system has to be represented in order to be successively modelled by an object-oriented approach, with particular reference to multiple hazards, poorly treated in literature.

7.2.2 Time Horizon of Events and Hazard Representation

With this representation, it is possible to follow the system during its lifespan (Fig. 7.3). A scenario-based approach needs to be adopted for the description of the system if subjected to hazards. Considering the two main components of the system as defined above (physical and social components), two quality indexes are defined:

- The quality index for the physical component is the value of the fragility of the system for the selected scenario (scenario-based fragility or SBF).
- The quality index of the social component needs to be defined starting from the well-known Life Quality Index (LQI), obtaining a modified LQI (MLQI).

With reference to Fig. 7.3, during its life, the system is subjected to discrete (e.g. earthquake, explosion, flooding) or continuous (e.g. aging) occurrences (events). The discrete events can be of two groups (Olmati et al., 2013):

1. High Probability- Low Consequence events, HP-LC events.
2. Low Probability- High Consequence events, LP-HC events (including black-swans).

It is expected that in case of LP-HC events the loss of system integrity is higher than for HP-LC events (Bontempi, 2005): the research on resilience should address just these situations.

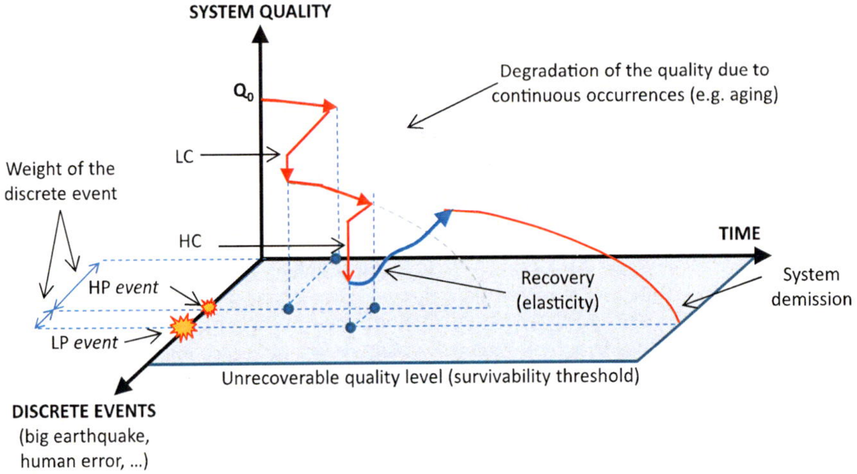

Fig. 7.3 Combined effect of discrete events and continuous deterioration

Considering the above, a series of issues arise regarding resilience-based design. The most important is maybe the choice of hazard scenarios that can influence resilience. The problem arises from the way to consider multi hazard scenarios. These can manifest in the following three different modalities (Petrini & Palmeri, 2012):

(i) Independent hazards, when different hazards affect the structure independently. For example, in the case of wind and earthquake hazard, they can be considered as independent of each other because no mutual interaction between the two hazards has the effect of modifying the intensity of the corresponding actions. These hazards can occur individually or simultaneously.
(ii) Interacting hazards, when the actions produced on a structure by different hazards are interdependent (e.g., wind and wave hazards, or wind and windborne debris hazards).
(iii) Hazard chains, when the effects of some hazards modify sequentially the effects of other hazards. For example, the actions on a structure due to windborne debris can damage the structural envelope and increases the vulnerability of the structure to strong winds. The same applies for fire hazard after earthquake.

These considerations serve for the successful implementation of decision representation at various levels, as shown in Sect. 7.2.3.

This is something common in structural design – that different hazards (thus, different loading schemes on the structure) have different design requirements. For example, in the case of high-rise buildings or long span bridges, and considering wind and earthquake loading, the first is the one that governs the structural design. Furthermore, optimizing for one hazard can have a counter effect on another. All these, not considering complications arising from common serviceability limits.

7.2.3 *Decisions and Data Driven Modelling*

Referring to a potential disaster of the system (significant loss of quality), its genesis is modelled by a Swiss Cheese Model (Reason, 1997). In the Swiss Cheese Model, the defences against failure are modelled as a series of barriers, represented as slices of cheese. The holes in the slices represent weaknesses in individual parts of the system and are continually varying in size and position across the slices. The system produces failures when a hole in each slice momentarily aligns, permitting (in Reason's words) "a trajectory of accident opportunity", so that a hazard passes through holes in all of the slices, leading to a failure or losses.

7.2.3.1 Representation of Decisions

One of the aspects that has a dominating influence on resilience is the political decision. Political decisions have a decisive impact on performance of a society, something that affects all elements of our technical infrastructure. In this paragraph, the losses refer specifically to the political decisions. The logical steps identified in Fig. 7.4, allow representing in an ideal way the losses computation.

The events affecting the system quality during its life, are described using the resilience framework (see Fig. 7.4). Referring as an example, to the political decisions, in resilience terms, five phases can be identified in a specific timespan of the system life specifically related to a discrete occurrence of events:

(i) historical
(ii) pre-event
(iii) during the event
(iv) aftermath of the event (recovery phase)
(v) long-run

The political decisions can influence all five above-mentioned phases as described below and in Fig. 7.5.

(i) Historical political decisions (e.g. political regime) influence the initial system quality Q_0. This parameter is related the initial system state. A high value of Q_0

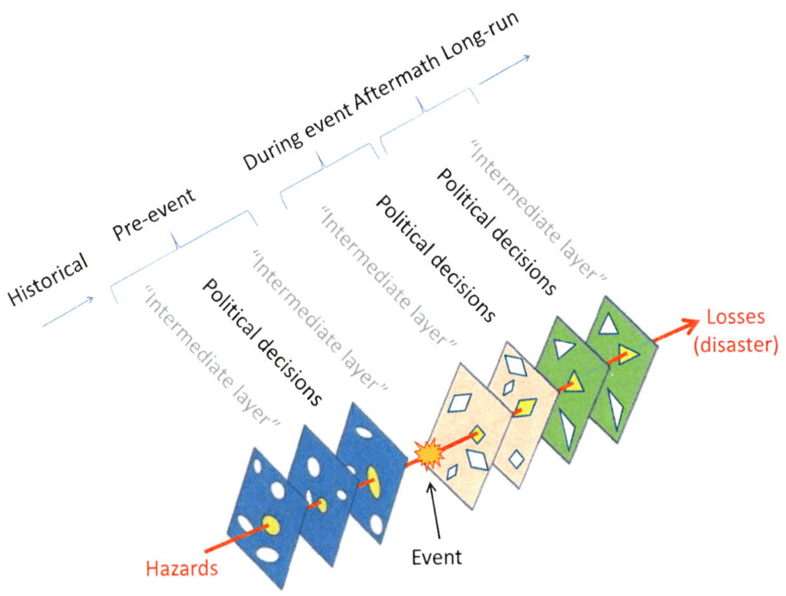

Fig. 7.4 Logical steps preliminary to the Losses computation. (Source: Adapted from Reason, 1997)

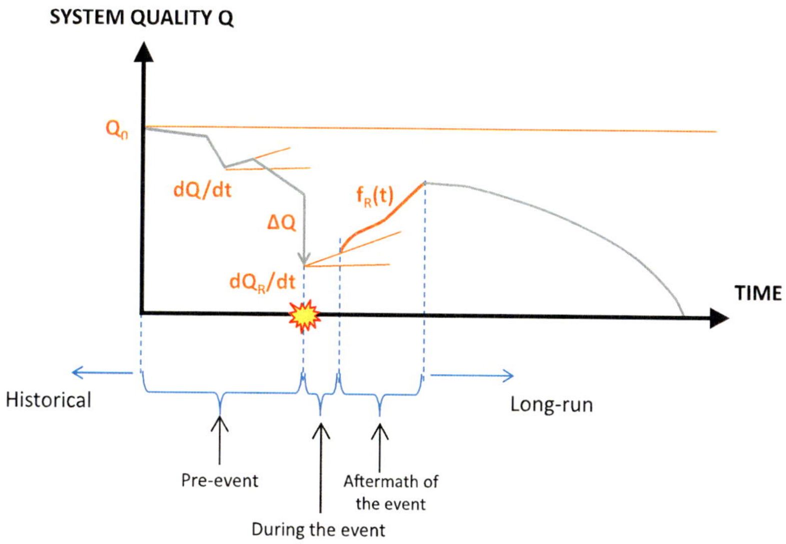

Fig. 7.5 System quality or functionality over time with different events and responses

indicates a "good" initial condition, and is an asset in the development of the different events occurring in the lifetime of the system.

(ii) Urban and social service planning is relevant for the pre-event system integrity and it is one of the factors determining the trend of system quality before the event (dQ/dt) and the amount of immediate losses (ΔQ). The first parameter is by a statistical correlation between the political decision and the quality trend, while the second parameter can be modelled by the system fragility.

(iii) Fast decisions are taken during the event or a disaster (on the basis of knowledge and experience from similar past events – if any – and the system properness knowledge). The political decisions in this phase influence the initial slope of the recovery phase (dQ_R/dt). This parameter can be modelled by a statistical correlation between the political decision and the quality trend.

(iv) Emergency plans and prioritization of recovery actions are relevant in the aftermath of the event (recovery phase). In addition, the declaration of a state of emergency can have a substantial short and long-term effect on the local economy. The actions in this phase influence the shape of the recovery function $f_R(t)$.

(v) Urban and social service re-planning on the basis of the consequences of the occurred event are relevant in the long run (influencing the losses and the recovery for the next discrete event).

Considering the above, the consequences of the political decisions can be modelled in the different phases as follow:

(i) Referring to some fixed typologies of historical political decisions, appropriate data-driven statistical relationships (mean value and standard deviation) will be defined between:

- the values of the human development index (HDI) – also considering the climate state of the system – and the initial value of the SBF of the individual system components.
- the values of the HDI and the value of the MLQI for modelling the social component.

(ii) Concerning the social component, appropriate factors can be defined in order to change the value of the MLQI in correspondence to a set of scenario-based pre-event political decision and evolution time ranges will be associated to each of them. Concerning the physical component, appropriate morphing modes (e.g. translation and rotation) can be defined for the fragilities of the system components in relation to a set of scenario-based pre-event political decision.

(iii) In order to model the consequences of the political decisions taken during the event appropriate data-driven statistical relationships (mean value and standard deviation) will be defined between the amount of initial resources investment and the MLQI or the decreasing rate of the SBF in time.

(iv) In order to model the consequences of the political decisions taken in the aftermath of the event (recovery phase) a class of recovery function shapes $f_R(t)$ will be defined for both the MLQI and the SBF as a function of the set of political decision previously defined.

(v) The consequences of the political decision can be classified in two groups:

(a) Direct consequences: are those affecting an element of the system which is directly interested by the political decision. An example of direct consequence is the one that we have in case of political decision regarding the maintenance of a critical infrastructure (e.g. a firefighters' station). Thus, an inadequate maintenance results in a significant increase in the system fragility to the hazard (defined as the probability of suffering a defined drop of the quality for a given intensity of the hazard), the absence of maintenance for some years is modelled by a slight increase in the fragility, while an appropriate maintenance results in an approximately constant fragility in time. An improvement action is modelled by a decrease in the fragility.

(b) Indirect consequences: are those affecting an element of the system that has been recognized to be dependent from another element affected by direct consequences. An example of indirect consequence is the one that we have when a building is subjected to a fire triggered by a big earthquake (hazard-chain scenario), when the nearest firefighter station has been seriously damaged by the same earthquake. The fragility of the building to the event is increased since the time intervention of the firefighters has been

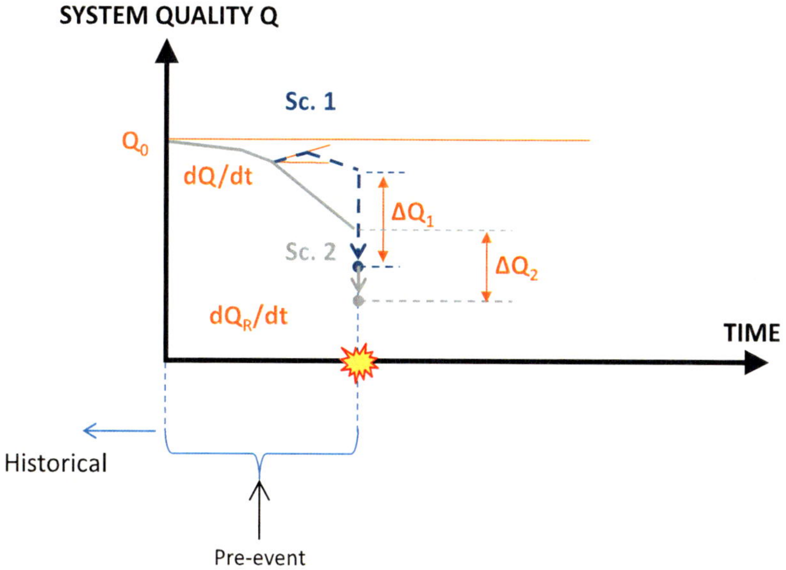

Fig. 7.6 System quality or functionality over time under different scenarios

increased. Indirect consequences are strongly related to the influence and interaction levels between the elements of the system as introduced above.

All the above-mentioned modelling strategies for the consequences of political decisions are applied in order to evaluate both direct and indirect consequences.

Just to make a simple example, as Fig. 7.6 suggests, political decisions can influence the slope dQ/dt. For example, if we consider two scenarios regarding the maintenance or improvement of an infrastructure (e.g. a bridge). Scenario 1 represents one with an improvement (continuous) at a certain point in the pre-event phase, while scenario 2 is the one with the normal deterioration over time. Obviously, the above mentioned scenarios are not exhaustive. When an extreme event occurs, the loss of quality ΔQ_1 will be different to ΔQ_2 and represents the quality excess due to the original political decision at the time of the event. Similar considerations can be made for the different phases of the time-line and appropriate metrics of system quality evolution can be obtained that can be quantified and inserted in planning processes (e.g. cost-benefit or cost-effectiveness analyses).

7.2.3.2 Data Collection for Data Driven Modelling

Data available from literature, statistical databases, empirical and heuristic sources, satellite earth observation and the internet will be used in order to set/calibrate the above mentioned models (Data-Driven Modelling approach/Big Data).

Data driven modelling is based on analysing the data about a system, and in particular finding connections between the system state variables (input, internal and output variables) without explicit knowledge of the physical behaviour of the system. These methods represent large advances on and include contributions from several fields.

1. Available data regarding the vulnerability of the elements of the system, the connections and correlations between elements, the service and efficiency level of them are used to establish the coupling and interaction levels between the system elements (see for example Arangio et al., 2013).
2. Historical data regarding the environment and the occurrence of different hazards are used to classify the correlations between different hazards in the categories defined above and to establish appropriate multi-hazard scenarios (including political decisions) for resilience assessment and to test the system behaviour.
3. Available data together with the data obtained by preliminary analyses describing the system behaviour under different political decisions, are used as a basis for defining a set of information-based political decisions and for quantifying the improvement of the system resilience due to the well-practice in the management of the system in the five phases defined above.
4. Appropriate methods for the data driven modelling of the system and the hazard will be identified from the field of artificial intelligence. Compared to conventional empirical modelling, a tangible and effective way to depict a system and its evolution is only by a distributed and embedded mathematical apparatus as proposed by soft-computing techniques. Soft computing is the effective state-of-the-art approach to artificial intelligence. By implementing soft computing methods, it is possible to exploit the tolerance to the imprecision and to the uncertainty to achieve tractability and robustness with little computational cost, being the human mind the model of reference. These methods of analysis are therefore suitable for the treatment of problems in which an exact solution does not exist or if exists it is a very complex one (Zadeh, 1997). In particular:

   (i) to manage the ambiguity and the uncertainty in probabilistic terms using fuzzy systems (systems that allow reproducing the approximate reasoning of the human mind);
   (ii) to handle spatial and temporal distributed characteristics with artificial neural networks (systems of elaboration of the information inspired by the biological neural network, characterized for their robustness and flexibility and for their ability to generalize) able for example to take advantage from satellite data and google maps for representing the system that can be properly calibrated by the using the available data and info collected using nowadays available techniques like evolutionary algorithms and other bio-inspired methods.

7.3 Antifragility Challenges and Risk Assessment

Very briefly (Taleb & Douandy, 2013) *fragility* is related to how a system suffers from the variability of its environment beyond a certain preset threshold while *antifragility* refers to when it benefits from this variability. In other words, systems range from *fragile* (degrading with stress), to *robust* (unchanged by stress), to *antifragile* (improving with stress).

Antifragility builds on a previous work of Taleb on Black Swans, very rare events that lie in the tails of distributions, and often beyond a specific sample range.

It is a common perception that Black Swans (and X-Events for this matter – see Casti, 2012) have changed the designers' perception after the shocking events of September 11 2001.

To reconnect with what stated before, simply put, fragility and antifragility respectively mean potential of harm or gain as a consequence to exposure to a volatility factor. This "something", as Taleb states (see, Keating, 2013), belongs to the extended disorder family (or Cluster): (i) uncertainty, (ii) variability, (iii) imperfect incomplete knowledge, (iv) chance, (v) chaos, (vi) volatility, (vii) disorder, (viii) entropy, (ix) time, (x) the unknown, (xi) randomness, (xiii) stressor, (xiv) error, (xv) dispersion of outcomes, (xvi) "unknowledge" (this one as an antonym for knowledge).

A challenge remains on how to quantify antifragility. Even though, fragility is rather easily measured (or better, compared) using metrics, and the use of fragility functions is common nowadays, this is not the case for antifragility.

Aven (2015), on the basis of Taleb's work, suggests using the notion of "asymmetry", that is, the idea that if a random effect has more upside effects that downside effects, is antifragile. Otherwise, it is fragile. The idea is to measure the harm induced by shock: if it gets higher as the intensity of the shock increases, the system can be considered as fragile. Otherwise (that is, if the harm is relatively low – what could be called a beneficiate to the system) the system is antifragile. This concept has dissimilarities with the conceptual idea that "robustness" lies between fragility and antifragility (in this case, somehow antifragility coincides with a "dynamic" robustness), however, it is a simple way to qualitatively describes something otherwise difficulty quantifiable.

Johnson and Gheorghe (2013) focus on the antifragility assessment of complex adaptive systems, and provide a case study on smart grid electrical systems, using a series of analytical criteria that characterize the system as fragile, robust or antifragile. In their case, the antifragility criteria met coincide with issues arising from positive outcomes of inducing stressors and learning from mistakes. This is not something uncommon. Since the start of the millennium, there have been attempts to induce (controlled) stressors in systems to assess their resilience. This is the case of Amazon GameDay project, with similar efforts from google and others (see Robbins et al., 2012). In this sense, a system can become antifragile, since, it grows stronger from each successive stressor, disturbance, and failure, in a "lessons learned" manner. The more frequently failure occurs, the more prepared the system and organization become to deal with it in a transparent and predictable manner (Tseitlin, 2013).

What should be clear is that antifragile design (and this is also the case for resilient based design) spans a wide area of topics, wider than commonly implemented methods for risk assessment and instruments, with decisions taken on the basis of scenarios and good judgement.

The profound understanding of a number of concepts, some common, some other borrowed from other fields (e.g. financial science, psychology) is necessary in order to pursue antifragility in the design, especially within a risk analysis framework. Issues and concepts that need to be considered in the era of antifragile design include:

- *Multi-hazard* (design). Natural hazard types have very different characteristics, in terms of the spatial and temporal scales they influence, hazard frequency and return period, and measures of intensity and impact (think for example earthquake and strong wind). The low frequency of the contemporary presence of such hazards is accounted for in structural limit state design (and this is the case for uncorrelated hazards). However, some hazards are a consequence of others (e.g. an earthquake may trigger landslides or fire, whereas a wildfire may increase the probability of future landslides – see Gill & Malamud, 2014). A "multi-hazard" design approach that pursues to identify all possible natural hazards and their interactions or interrelationships is nowadays essential.
- *Black swan*. This term was proposed by Taleb (2007). Other similar terms have been introduced in the recent years ("known unknowns", X-Events, etc.) to describe rare events, with extreme impact, and retrospective.
- *Halo effect*. A cognitive process in which the global evaluation of something or someone can influence one's response to other attributes or the impression of one attribute shapes the impression of another independent attribute. In risk analysis it can lead to wrong judgement.
- *Gambler's fallacy*. The (mistaken) belief that, if something happens more frequently than normal during some period, it will happen less frequently in the future, or that, if something happens less frequently than normal during some period, it will happen more frequently in the future (Tune, 1964). It arises from the erroneous belief that small samples must be representative of the larger population. As in the previous case, it can lead to misjudgment.
- *Synchronicity*. A term coined by Jung (1960) to express a concept about the acausal connection of two or more psycho-physic phenomena, that is, the "timing together" of otherwise "unrelated" events. In risk analysis, can be used to develop a broad view of phenomena otherwise standardized in clusters.
- *Apophenia*. Although the term has its basis in psychiatry (used to describe early stages of schizophrenia), it is nowadays intended as the experience of seeing meaningful patterns or connections in random or meaningless data. In risk analysis it is linked with statistical errors, while a positive effect, can be the added resourcefulness in scenario planning.

- *Dependability*. It is concisely defined as the grade of confidence on the safety and on the performance of a system (see Sgambi et al., 2012 for a dependability framework in the civil engineering field). This is a qualitative definition that comprehensively accounts for several properties, which, even though interconnected, can be examined separately. Robustness is a dependability attribute.
- *Bias* (see also *selection bias*). In risk analysis, it is important to select unbiased data, group, people, etc. Experts are often biased towards expected results and (non deliberately) drive their results. See also *cherry picking*, *suppressing evidence*, or the *fallacy of incomplete evidence*
- *Self-deception (*also *subjective validation)*. The art of convincing and validating oneself. Can prove to be negative in risk brainstorming activities.
- *Swarm effect*. It is the collective behavior that emerges from a group of social insects (or humans, for what matters). Through this effect, people may group together, share the same influences and drive towards the same goal or beliefs. The swarm effect has a strong effect on political decisions that influence infrastructure planning and maintenance, thus, influence resilience.
- *Retrospective*. It is an attribute of black swans, and helps explain facts after their occurrence. To make a simple case, everyone after September 11 2001 considers the possibility of an airplane impact on strategic structures (something less rare in the 60s and 70s – considering also nuclear plants construction) but faded afterwards.
- *Mindfulness* (collective). Another concept from human behavior, essentially to "live as if you were living for the second time and have acted wrongly the first time as you are about to act now" (see also *mindful management* – Weick & Sutcliffe, 2007).

Some of the above mentioned concepts date back to several years. However, their non-consideration in modern risk assessment lead to ignoring critical aspects and may compromise the entire process.

7.4 Conclusions

The authors' intention is to review recent developments, together with corroborated research, focusing on new trends in the resilience-based design, focusing on the political decisions behind it. In fact, major events have major effects on a community scale and drive the public opinion towards new demands, and as a consequence, influence policy makers and politicians. This is how concepts such as those treated in this study surfaced and became popular research topics. Resilience-based design became a hot topic in the last 15 years, after major catastrophic events with extreme impact occurred on a community scale. Nevertheless, issues remain to be solved. There are still relativelly few resilience standards, and even resilience metrics are somehow difficult to implement. The innovative concept of antifragility can have a major influence the resilience-based design.

The modelling of the influence of political decisions on community resilience is nowadays possible, through the exploration of new methods (e.g. big date) and the system (and decision) representation in a formal way. However, at the moment, the connection between political decisions and antifragility is trivial, and possible developments will become clear only when research objectives on antifragility and the built environment are fully understood.

Acknowledgements This study presents methods, considerations and results, developed in the last years by the authors. It is also based on considerations by the authors during and after an Informal Meeting on Resilience organised by Prof. Franco Bontempi (Sapienza University of Rome) and by the late Prof. Alessandro de Stefano (Politecnico di Torino) in Rome on July 2–3 2014. At the time this study took place Konstantinos Gkoumas was employed by the Sapienza University of Rome.

References

American Psychological Association. (n.d.). APA dictionary of psychology. Retrieved June 15, 2022, from https://dictionary.apa.org/resilience

Aven, T. (2015). The concept of antifragility and its implications for the practice of risk analysis. *Risk Analysis: An International Journal, 35*(3), 476–483.

Arangio, S., Calò, F., Di Mauro, M., Bonano, M., Marsella, M., & Manunta, M. (2013). An application of the SBAS-DInSAR technique for the assessment of structural damage in the city of Rome. *Structure & Infrastructure Engineering, 10*(11), 1469–1483.

Asprone, D., Cavallaro, M., Latora, V., Manfredi, G., & Nicosia, V. (2013) Urban network resilience analysis in case of earthquakes. In *Proceedings of the 11th international conference on structural safety & reliability*. , June 16–20.

Bontempi, F. (2005). Frameworks for structural analysis. In B. H. V. Topping (Ed.), *Innovation in civil and structural engineering computing* (pp. 1–24). Saxe-Coburg Publications.

Bontempi, F., Gkoumas, K., & Arangio, S. (2008). Systemic approach for the maintenance of complex structural systems. *Structure & Infrastructure Engineering – Maintenance, Management, Life-Cycle Design & Performance, 4*(2), 77–94.

Bosher, L. (Ed.). (2008). *Hazards and the built environment: Attaining built-in resilience*. Taylor and Francis.

Bruneau, M., Chang, S., Eguchi, R., Lee, G., O'Rourke, T., Reinhorn, A. M., Shinozuka, M., Tierney, K., Wallace, W., & Winterfelt, D. V. (2003). A framework to quantitatively assess and enhance the seismic resilience of communities. *Earthquake Spectra, 19*(4), 733–752.

Calvino, I. (1974). Invisible cities. Paperback, 165 pages Published 1974 by Harcourt Brace & Company.

Casti, J. L. (2012). *X-Events: The Collapse of Everything*. Harper Collins.

Castoriadis, C. (1987). *The imaginary institution of society*. The MIT Press.

Cavalieri, F., Franchin, P., Gehl, P., & Khazai, B. (2012). Quantitative assessment of social losses based on physical damage and interaction with infrastructural systems. *Earthquake Engineering & Structural Dynamics, 41*(11), 1569–1589.

Cimellaro, G. P. (2016). *Urban resilience for emergency response and recovery: Fundamental concepts and applications*. Springer.

Cimellaro, G. P., Reinhorn, A. M., & Bruneau, M. (2010). Framework for analytical quantification of disaster resilience. *Engineering Structures, 32*(11), 3639–3649.

Coaffee, J. (2008). Risk, resilience, and environmentally sustainable cities. *Energy Policy, 36*(12), 4633–4638.

Davoudi, S., Brooks, E., & Mehmood, A. (2013). Evolutionary resilience and strategies for climate adaptation. *Planning Practice & Research, 28*, 307–322.

Franchin, P., & Cavalieri, F. (2015). Probabilistic assessment of civil infrastructure resilience to earthquakes. *Computer-Aided Civil and Infrastructure Engineering, 30*(7), 583–600.

Gill, J. C., & Malamud, B. D. (2014). Reviewing and visualizing the interactions of natural hazards. *Reviews of Geophysics, 52*(4), 680–722.

Gunderson, L. H. (2000). Ecological Resilience – In theory and application. *Annual Review of Ecology and Systematics, 31*, 425–439.

Gunderson, L. H., & Holling, C. S. (2002). *Panarchy: Understanding Transformations in Human and Natural Systems*. Island Press.

Holling, C. S. (1973). Resilience and stability of ecological systems. *Annual Review of Ecology and Systematics, 4*, 1–23.

Johnson, J., & Gheorghe, A. V. (2013). Antifragility analysis and measurement framework for systems of systems. *International Journal of Disaster Risk Science, 4*(4), 159–168.

Jung, C. G. (1960). *Synchronicity. An Acausal Connecting Principle*. Princeton University Press.

Keating, C. (2013). Antifragile: How to live in a world we don't understand. *Quantitative Finance, 11*(13), 1693–1695.

Lindell, M. K. (2010). Built-in resilience. *Nature Geoscience, 3*(11), 739–740.

Maruyama, H., Legaspi, R., Minami, K., & Yamagata, Y. (2014). General Resilience: Taxonomy and Strategies. In *Proceedings of the International Conference and Utility Exhibition 2014 on Green Energy for Sustainable Development* (ICUE).

MCEER – Multidisciplinary Center for Earthquake Engineering Research. (2006). MCEER's resilience framework. Retrieved September 6, 2016 from: http://mceer.buffalo.edu/research/resilience

Meerow, S., & Newell, J. P. (2015). Resilience and complexity: A bibliometric review and prospects for industrial ecology. *Journal of Industrial Ecology, 19*(2), 236–251.

Mostashari, A., & Sussman, J. (2009). A framework for analysis, design and management of complex large-scale interconnected open sociotechnical systems. *International Journal of Decision Support System Technology, 1*(2), 52–68.

Olmati, P., Gkoumas, K., Brando, F., & Cao, L. (2013). Consequence-based robustness assessment of a steel truss bridge. *Steel and Composite Structures – An International Journal, 14*(4), 379–395.

Ortenzi, M., Petrini, F., Bontempi, F., & Giuliani, L. (2013, June 16–20). RISE: A method for the design of resilient infrastructures and structures against emergencies. In *Proceedings of the 11th International Conference on Structural Safety & Reliability*. New York.

Pearson, C., & Delatte, N. (2005). Ronan Point apartment tower collapse and its effect on building codes. *Journal of Performance of Constructed Facilities, 19*(2), 172–177.

Perrow, C. (1999). *Normal accidents: Living with high-risk technologies*. Princeton University Press.

Petrini, F., & Palmeri, A. (2012, July 8–12). Performance-based design of bridge structures subjected to multiple hazards: a review. In *Proceedings of the sixth international conference on bridge maintenance, safety and management,*. Stresa, Lake Maggiore, Italy.

Renschler, C., Frazier, A., Arendt, L., Cimellaro, G. P., Reinhorn, A. M., & Bruneau, M. (2010, July 25–29). Developing the "PEOPLES" resilience framework for defining and meas-uring disaster resilience at the community scale. In *Proceedings of the 9th US National and 10th Canadian Conference on Earthquake Engineering*, Toronto, Canada.

Reason, J. (1997). *Managing the risks of organizational accidents*. Ashgate. ISBN 1840141042.

Robbins, J., Krishnan, K., Allspaw, J., & Limoncelli, T. (2012). Resilience engineering: Learning to embrace failure. *Communications of the ACM, 55*(11), 40–47.

Salthe, N. S. (1985). *Evolving hierarchical systems*. Columbia University Press.

Sgambi, L., Gkoumas, K., & Bontempi, F. (2012). Genetic algorithms for the dependability assurance in the design of a long span suspension bridge. *Computer-Aided Civil and Infrastructure Engineering, 27*(9), 655–675.

Taleb, N. N. (2007). *The Black Swan: The impact of the highly improbable* (1st ed.). Penguin.

Taleb, N. N. (2012). *Antifragile: Things that gain from disorder*. Penguin.
Taleb, N. N., & Douandy, R. (2013). Mathematical definition, mapping, and detection of (anti) fragility. *Quantitative Finance, 13*(11), 1677–1689.
Tseitlin, A. (2013). The antifragile organization. *Communications of the ACM, 56*(8), 40–44.
Tune, G. S. (1964). Response preferences: A review of some relevant literature. *Psychological Bulletin, 61*, 286–302.
Weick, K. E., & Sutcliffe, K. M. (2007). *Managing the unexpected: Resilient performance in an age of uncertainty* (2nd ed.). Wiley.
Zadeh, L. (1997). What is soft computing. *Soft Computing, 1*(1), 1.

Chapter 8
An Integrated Methodological Framework to Assess Urban Resilience

Julia Martínez-Fernández, Miguel Angel Esteve-Selma, Isabel Banos-Gonzalez, and Noelia Guaita-García

Abstract The assessment of the urban resilience should be tackled with a systemic perspective that enables an integrated analysis of the environmental, social, economic and institutional factors and their interactions characterizing urban and other complex socio-ecological systems. Here we propose an integrated framework for such assessment with the following key components: (i) The hierarchical definition of resilience objectives and indicators. (ii) A dynamic system model taking into account the key socio-economic and environment factors and their interactions, in which resilience indicators are integrated. (iii) The assessment of model potential sources of uncertainty and their impact on model outputs. (iv) The analysis of vulnerabilities to exogenous drivers (scenario analysis) and the exploration of available management and planning options (policy assessment). (v) A multi-criteria procedure, in which indicators, resilience thresholds, model outputs and scenario and policy analysis are integrated to guide decisions for an improved urban resilience. The whole framework integrates a participative approach, mainly for the initial and final steps.

Keywords Dynamic models · Policy assessment · Socio-ecological systems · Sustainability indicators

J. Martínez-Fernández (✉)
Association for Sustainability and Progress of Societies, ASYPS, New Water Culture Foundation, Zaragoza, Spain
e-mail: julia@fnca.eu

M. A. Esteve-Selma
Department of Ecology and Hydrology, University of Murcia, Murcia, Spain
e-mail: maesteve@um.es

I. Banos-Gonzalez
Department of Experimental Sciences Didactic, University of Murcia, Murcia, Spain
e-mail: ibbg1@um.es

N. Guaita-García
Department of Life Sciences, University of Alcala, Alcalá de Henares, Madrid, Spain
e-mail: noelia.guaita@uah.es

O. F. González Castillo et al. (eds.), *Urban Resilience: Methodologies, Tools and Evaluation*, Resilient Cities, https://doi.org/10.1007/978-3-031-07586-5_8

8.1 Introduction

Urban areas can be considered a particular type of socio-ecological systems, where social, economic and environmental factors interact in a nonlinear fashion characterized by their reinforcing mechanisms. As other complex system, urban areas can face systemic changes, arising either from an external hazard event or from gradual endogenous change (Filatova & Polhill, 2012). The analysis of this dynamics, basic to understand the resilience of urban systems, requires a holistic approach, as for other socio-ecological systems (Lacitignola et al., 2007; Hodbod & Adger, 2014).

The integrated approach needed for more resilient urban systems should also consider wide time and spatial scales, to address the connections between urban systems with other surrounding socio-ecological systems. As it has been emphasised (Suárez et al., 2016), is it necessary not only to create more resilient cities, but also to reduce their environmental and social impacts on natural ecosystems and agricultural landscapes at broader scales. For example, urban sprawl constitutes one of the main factors driving the disappearance of traditional agrosystems and cultural landscapes around cities.

However, despite the increasing acknowledgement about the need for such holistic, systemic approaches, the application of integrated perspectives in urban and other socio-ecological systems are less frequent that desirable. Among the difficulties behind this, it should be emphasised (i) the need for a new conceptual perspective concerning the relationships between science and the management of real systems and (ii) the lack of tools to manage the inherent complexity of such systems. In the following sections, these two difficulties are further discussed.

8.1.1 *The Need for a New Conceptual Perspective*

Regarding the first difficulty, in recent years new conceptual approaches have been proposed; among them, science for sustainability or the post-normal science, as opposed to the discipline-oriented view of the positivist science. Table 8.1 synthetizes, according to Haag and Kaupenjohann (2001), the main differences between these two perspectives.

Following the perspectives emerging from the science for sustainability and post-normal science, in the approach proposed here there are two inter-connected but clearly distinctive components: The first one refers to the description of the system, making use of all necessary data and models. This description should be dynamic, to tackle not only the present state and problems but also their potential future evolution and should also be integrative, so synergies and trade-offs between different factors, objectives and actions are fully considered.

The second component refers to the definition of objectives, criteria and valuation procedures to assess the vulnerabilities of urban systems and to select the most appropriate policy measures among a set of alternatives and options, on the basis of

Table 8.1 Main differences between the positivist science and the post-normal science (Haag & Kaupenjohann, 2001)

Normal, positivist science	Post-normal science
Well defined theoretical systems	Ill-defined real problems entangled in complex ecological-socioeconomic systems. Non-equivalent descriptions are possible. Role of stakeholders in problem/system definition
Universal	Specific, unique systems
Independent of problem issues	Problem-driven. Modelling for management
Context-free	Context-sensitive
Scientific disciplines, reductionist approach	Trans-disciplinary, holistic approach
Systems to be studied: abstract, idealized	Systems to be studied: real cases
Very limited consideration (technical) or complete exclusion of uncertainty. Validation/quality control by a close community of experts	Deep consideration of different types of uncertainty, some of which are non-reductible. Extended peer community including stakeholders
Straightforward use in management. Frequently, using optimization models, assuming linear relationships and a single valuation criterium, usually defined by a close group of experts.	Increased relevance of values. Valuation procedures involving very different stakeholders. Scientific data and model results as inputs for valuation and decisión-making processes, dealing with multiple criteria, alternative valuations, incommensurability and uncertainty

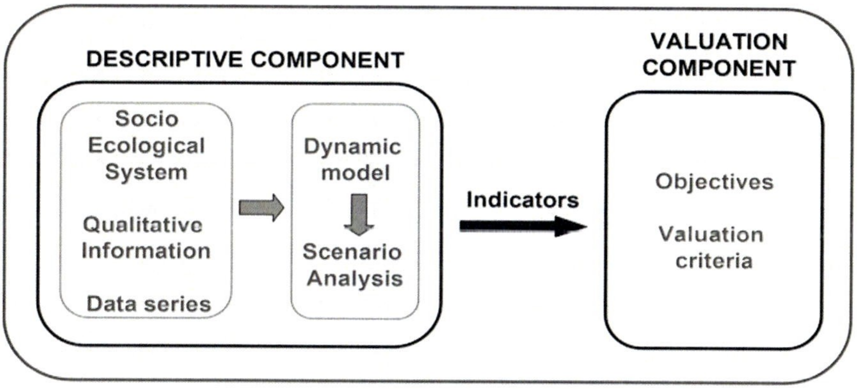

Fig. 8.1 General components for the assessment of system resilience

the best available knowledge provided by the descriptive component. The second component should answer two main questions: (i) which are the objectives to be achieved and the criteria to be considered for more resilient urban systems and (ii) which are the relative contributions of different policy options to urban resilience and which ones should be considered more beneficial to be implemented. Figure 8.1 shows the relationships between both components.

8.1.2 Integrated Tools to Assess Urban Resilience

In this chapter we propose an overall methodological framework which integrates the following components and tools:

1. The hierarchical definition of resilience objectives, along with their indicators and thresholds
2. A dynamic simulation model, where the above indicators are integrated
3. The assessment of policy options and the analysis of vulnerabilities to external scenarios
4. An uncertainty assessment concerning system behaviour and model outcomes
5. A procedure to assist decisions for an improved urban resilience

Figure 8.2 presents the relationships among the basic components. This overall approach addresses multiple purposes, including the capacity of anticipation, which is also a key property of resilient urban systems and societies (Khazai et al., 2015). The participation of involved agents, particularly in the first and last stages, is essential.

In the following sections these components and tools are described with more detail and some application examples are given.

8.2 Definition of Indicators. The IDIS Approach

8.2.1 Indicators as Tools for an Improved Resilience

One of the difficulties for an adequate understanding of resilience in urban and other socio-ecological systems is the existence of an overhelming amount of information referred to a wide number of aspects, frequently very heterogeneous in terms of

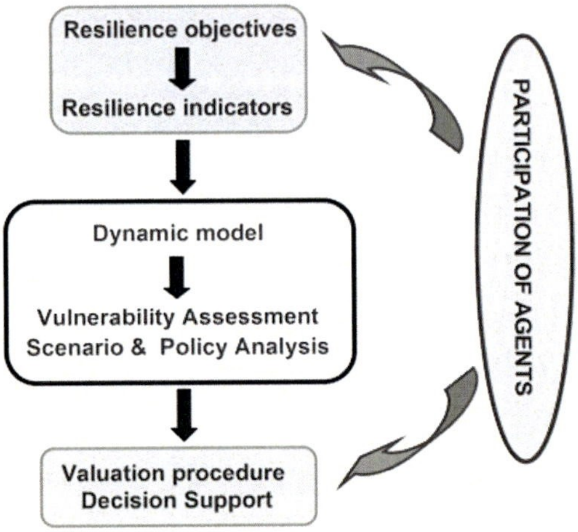

Fig. 8.2 Overall methodological framework to assess urban resilience

level of detail, spatial and temporal scales and other basic properties. To understand the multi-dimensional nature of resilience, which is a key component of sustainable systems, requires tools helping to reduce and organise the relevant information, so it can be transformed into useful knowledge. Indicators allow to monitor and assess key aspects of urban resilience in a quantitative way, to carry out comparative analysis and to provide relevant knowledge in a non-technical language to policy makers, managers, stakeholders and the general public for the decisión taking processes. Indicators should be relevant concerning the aspects they appoint to, sensible to changes, easily computed and understood, useful for communication and with a mínimum of overlap with other indicators (Adriaanse, 1993; Bell & Morse, 2008).

In recent years an increasing effort is devoted to the development of indicators of urban resilience (see for example Gonçalves & Marques da Costa, 2013; Khazai et al., 2015). Most of urban resilience indicators are directly connected to risk assessment, whereas there are few examples of indicators not linked to specific risks. These more general indicators refer mainly to energy, water and urban drainage systems (Suárez et al., 2016).

However, it is increasingly claimed that indicators just have a moderate weight on the adoption and assessment of sustainable and resilient policies and practices (Reed et al., 2006; Moldan et al., 2012). Among the limitations of conventional catalogues of indicators, we highlight:

- Its static dimension, which reduces the possibility of considering the synergies and trade-offs between indicators and the assessment of their future developments under different management options.
- Its lack of significance for each specific case of study.
- The "top-down" approach, which reduces the involvement and responsibility of different agents in monitoring objectives through such indicators
- The frequent absence of thresholds to determine whether the changes in the indicators are acceptable or not (Moldan et al., 2012).

The overall framework proposed in this chapter overcome these limitations. One crucial point is the selection of indicators, for which we propose the methodological approach outlined below.

8.2.2 The IDIS Hierarchical Approach

In the IDIS (Dynamic Integration of Indicators) hierarchical approach the following procedure is followed:

- Establishment of the overall resilience goals to be achieved.
- Identification of the resilience dimensions or components of the concerned system
- For each dimension, some specific objectives are defined.
- For each specific objective, it is necessary to formulate some strategic questions, relevant for the policy making process, to be answered. The answers to these

strategic questions determine whether the system is moving towards achieving the specific objective.

- Finally, to answer each strategic question, one or few relevant indicators are identified.

Figure 8.3 graphically describes the IDIS approach.

The final system of indicators derives from the objectives to be achieved, it is specific for each system, applies an integrative perspective and includes only the required indicators, avoiding redundance or indicators not connected with specific objectives. This approach also aims at carefully selecting a restricted number of indicators, since a limited and manageable number of indicators creates a more useful tool than a large number of unselected ones (Lancker & Nijkamp, 2000).

The participation of agents contributes to build confidence on the final indicators system and to increase the corresponsibility of policy makers and stakeholders in the application of the indicators. Moreover, a participatory definition of indicators also contributes to the corresponsibility in the achievement of the objectives, since an agreement on the diagnosis – from which an indicators system constitute one of the components – do not ensure but facilitates an agreement on the potential solutions to the identified problems.

The hierarchical connections between objectives, strategic questions and indicators help to define indicator systems that are truly consistent with the overall resilience goals to be achieved. For instance, efficiency indicators are very important, but an indicators system mainly based on these type of relative indicators (such as per capita or other per unit indicators) might improve while, at the same time, the system is moving away from resilience. This can be illustrated with the per capita emissions in Fuerteventura (Canary Islands), for which it is expected an improvement between 2012 and 2025, despite the rises in the consumption of resources and emissions in absolute terms (Banos-González et al., 2016). This can be explained by

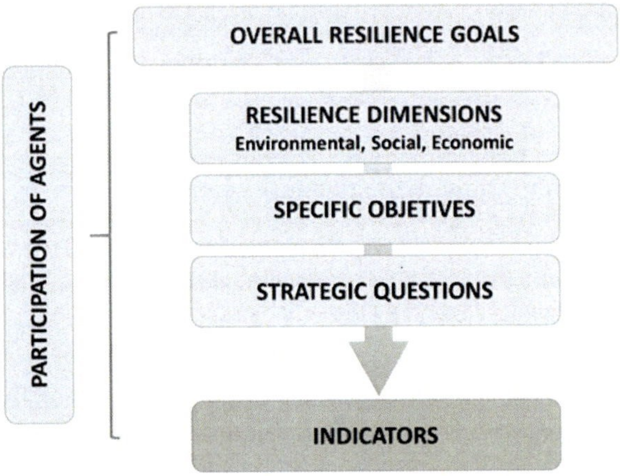

Fig. 8.3 Hierarchical approach for the definition of sustainability indicators

the even-higher expected increase in the total population over that period. Obviously, this does not imply that if more tourists come to the island, more resilient and sustainable Fuerteventura will be. This illustrates how some relative indicators, particularly many efficiency indicators, do not always give sound information, when considered alone. Therefore, these efficiency indicators and their changes over time should be taken with caution (Hanley et al., 2009), to avoid misunderstandings and errors in the diagnosis (Figge & Hahn, 2004; Mori & Christodoulou, 2012).

8.2.3 Resilience Thresholds

In terms of interpretation of the indicators and in order to represent a useful tool for the decision processes, a quantitative notion of what is acceptable for sustainability or for resilience – a threshold – is needed (Rodríguez-Rodríguez and Martínez-Vega, 2012; Banos-González et al., 2016). Without thresholds, indicators can describe, but are less useful as valuation tools to help decisions. Thresholds allow to track not only the direction and magnitude of changes but also to determine whether such changes are acceptable or not in terms of resilience or sustainability (Lancker & Nijkamp, 2000; Moldan et al., 2012; Proelss & Houghton, 2012). Indeed, the notion of threshold is key for an operative application of the concept of resilience (Gonçalves & Marques da Costa, 2013).

How can resilience thresholds be identified? A threshold may be a background value or it can be a meaningful reference value related to the irreversibility of the system. Depending on the nature of the indicator, threshold values can be provided by mandatory legal standards, guidelines from different institutions, benchmarking (best practices and experiences from other sites) and reference values taken from historical values of the system.

8.2.4 The Case of Galapagos Islands

The application of the IDIS approach to develop the water indicators system in the Galapagos Islands (SIAG, Martínez-Fernández et al., 2016) allows to discuss some of the limitations of conventional indicator catalogues, as explained below.

Static Versus Dynamic Indicators The static catalogues of indicators cannot take into account the interactions and trade-ofs among indicators, by which the improvement in some indicators may cause a worsening in others (Banos-González et al., 2016). Therefore, it is important to integrate the indicators within a dynamic model (Vidal-Legaz et al., 2013; Liu et al., 2014; Banos-González et al., 2015). The SIAG indicators are being integrated into a dynamic model covering the key dimensions for sustainability and resilience of the Galapagos Islands.

Need of Context-Specificity The SIAG includes several indicators regarding the resilience to climate change in the urban areas of Galapagos, characterized by a very arid climate. One of these indicators is the proportion of houses and other buildings which have devices for rainwater collection, a traditional water system which is being lost during the last decades. This system of water supply has other advantages, as the provision of high quality water for the most basic human needs with no or very low input of energy. However, other works have included as indicator of poverty in Galapagos the existence of water supply systems different to the public network (Granda Leon et al., 2013), since this is one of the indicators of poverty being applied in the continental Ecuador. This exemplifies how the direct translation of indicators to other areas can be inadequate to specific contexts, such as the Galapagos Islands, where rainwater collection is not related to poverty and, in fact, it should be promoted to increase Galapagos resilience to climate change. Moreover, the consideration of rainwater collection as an indication of poverty contributes to a negative perception of this device (Guyot-Tephany et al., 2013) and therefore is counterproductive to increase the overall resilience of the urban areas in Galapagos.

Participatory Versus Top-Down Approach The implication of agents are crucial in the development of urban resilience indicators (Khazai et al., 2015). A top-down, non-participatory approach, does not facilitate the co-responsibility of all agents in the effective application and follow-up of the indicators. In the case of SIAG, water managers and stakeholders have participated in its development, which has contributed to improve the initial proposal and to increase the interest of involved agents on an effective application of the indicators system.

Need of Thresholds A frequent weakness of many catalogues of indicators is the lack of reference values. Without thresholds, the indicators can describe, but are less useful as valuating tools to help decisions. In addition to monitor the direction and magnitude of change, thresholds allow us to determine whether such change is acceptable or not regarding resilience (Lancker & Nijkamp, 2000; Moldan et al., 2012; Proelss & Houghton, 2012). In the case of SIAG, a threshold was identified for each of the 34 indicators, what allowed to get some measure of the distance to goal and therefore to better prioritize the actions to be taken.

8.3 Dynamic Simulation Models

8.3.1 The System Dynamics Approach

System dynamics models (SD) allow us to understand the structure and behaviour of complex systems, by means of the causal relationships, feedback loops, delays and other processes of the system (Kampmann & Oliva, 2008; Li et al., 2012; Martínez-Moyano & Richardson, 2013). Negative feedback loops, which tend to absorb disturbances and maintain the overall behaviour within certain ranges, are also essential features for the resilience of socio-ecological systems. These feedbacks constitute, therefore, an important core urban resilience factor (Suárez et al., 2016).

The application of system dynamics modelling tools allow to facilitate the comprehension of complex systems (Martínez-Moyano & Richardson, 2013; Kelly et al., 2013) aimed at generating useful information for decision-making (Jakeman & Letcher, 2003; Voinov & Shugart, 2013). Another important feature of SD is its context-specific approach. Context-specific or context-adapted models are needed to be able of addressing the concrete problems, challenges and needs of real systems and, therefore, to provide proper solutions. Dynamic system models are particularly appropriate to visualize the overall system, to consider a long run perspective, to present factors and relationships in a transparent way and to integrate resilience indicators. All this make dynamic models valuable tools for a participatory management, helping in the communication among the scientific-technical, management and social agents' sides.

The modelling process involves several stages (Fig. 8.4): conceptualisation, formulation of model equations and calibration, in which the model is iteratively improved through calibration against the observed data of main variables. Finally, the model is tested by means of structural tests (Barlas, 1996), including dimensional consistence tests, sensitive analysis and extreme condition tests. After successful testing, the model is applied to define and assess the expected effects of different policies and scenarios. Figure 8.4 shows how these methodological steps are related.

In the next section the role of dynamic models is exemplified with the case of the periurban agro-ecosystem of Murcia, which has an essential role on the climatic

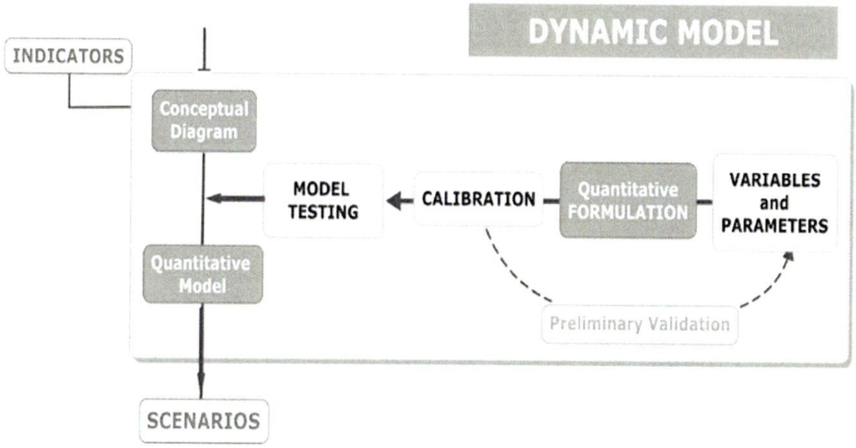

Fig. 8.4 Methodological steps to develop dynamic models

comfort of the city and is therefore essential to preserve the resilience of Murcia to climate change.

8.3.2 The Case of Murcia City and Its Periurban Agro-ecosystem

Dynamic models can help to better understand and improve the resilience of urban systems from an integral perspective. This is illustrated in the case of the city of Murcia (Southeastern Spain) and its periurban agro-ecosystem, called "huerta", a type of Mediterranean traditional irrigated land. Ancient elements like the irrigation ditches, dating from before 1000 AD, exist alongside modern elements associated to urban development. This irrigated agro-ecosystem plays a key role to substantially moderate the urban heat island effect of the city. The available regionalised climate change scenarios point to an increase in the intensity and frequency of heat waves, which will seriously impact on the public health and the thermal comfort of the city. Therefore, the preservation of the Huerta agro-ecosystem around Murcia is essential to improve the resilience of this urban system to the ongoing climate. Moreover, this traditional agro-landscape, called Huerta, has important environmental and cultural values.

Nevertheless, the preservation of this agrosystem is seriously threatened. In order to assess the factors involved in the progressive loss of this agroecosystem, it has been developed a dynamic system model (Fig. 8.5). The model takes into account the area occupied by the Huerta agro-ecosystem, the area occupied by new irrigated lands, located outside the river valley, the number of landowners, the average farm size, the population, the amount and quality of water resources used for irrigation and the profitability index. Several factors contribute to the decreasing profitability of the Huerta agro-ecosystem. The main factor is the reduction in the average farm size, caused by the permanent increase in the number of landowners in Huerta de Murcia. The consequent reduction in the average farm size strongly affects the profitability of Huerta. The scarcity and low quality of water resources used for irrigation also affects the profitability. Figure 8.5 shows a simplified diagram of the model. More details can be found in Martínez-Fernández et al. (2013).

In 1932 the Huerta de Murcia had 13,500 ha, area which progressively decreases to around 11,500 ha in 1995 (Fig. 8.6a). By the end of the period, around 15% of the initial area of Huerta had been lost, value which doubles the proportion of high quality soils lost in Spain (Comisión de las Comunidades Europeas, 1992).

One of the reasons for the loss of Huerta is the increase in total population by around 150% (Fig. 8.6b). Population demands land uptake for infrastructures and especially for residential uses. The other reason for the loss of huerta is the reduction of profitability due to the increase in the number of landowners (Fig. 8.7a), leading to the decrease in the average size per farm (Fig. 8.7b).

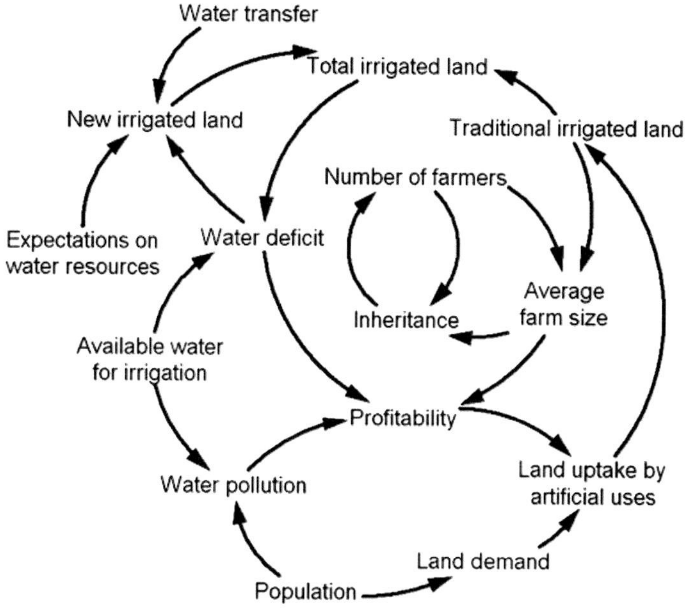

Fig. 8.5 Simplified diagram of the Huerta dynamic model

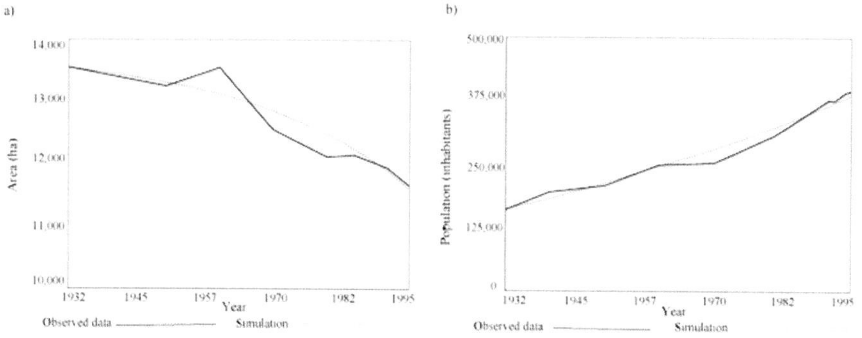

Fig. 8.6 Observed data and simulation results for (**a**) traditional irrigated lands around the city of Murcia, (**b**) total population

Under the base trend scenario, the traditional irrigated land would be lost between 1995 and 2025 at a rate which doubles the average annual land uptake in the previous 30-years period.

However, which has been the actual behaviour of the system after 1995? Results obtained show that the area of traditional irrigated land expected under the Base trend scenario by year 2025, has actually been reached 18 years earlier. One key factor for such acceleration is the new municipal land use policy implemented after 1995, which has shifted from weak or no controls on land uptake to an even worse

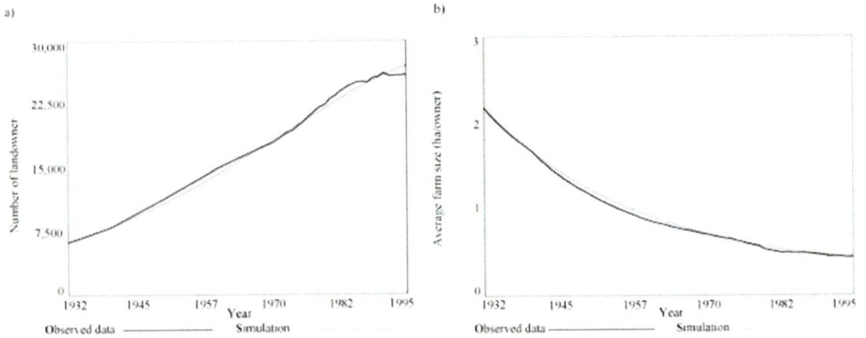

Fig. 8.7 Observed data and simulation results for (**a**) number of landowners of traditional irrigated lands, (**b**) average farm size

situation characterised by the successive reduction in the legally preserved area of the Huerta. Between 1998 and 2001 the legally preserved area of the Huerta was reduced by around 38%. Although the change in legal status from preserved to buildable area does not imply an immediate transformation into urban use, the expectations about this potential future urban population might be the key factor for the observed accelerated loss of the Huerta. To assess this, we have defined and implemented a new scenario corresponding to the actual trend 1995–2010 with two main aims: (i) to test whether the changes in the municipal land plan can explain the observed loss of the Huerta in 1995–2010 and (ii) to test whether the dynamic model, developed and calibrated with data from the 1932–1995 period, can resemble the observed trends in the 1995–2010 period without further calibration. The potential effect of the municipal land plan as been implemented by including the expected increase in population associated to the successive changes in the municipal land plan for years 1995, 2001 and 2005, taking into account the official values in each plan of buildable area, floor area ratio, average number of inhabitants per home and the time period required to reach the saturation of such buildable area. The area of the Huerta, number of landowners, population, average farm size and water quality index were updated for the period 1995–2010 for comparison purposes with simulation runs, but no new calibration against observed data was performed.

Results show that the actual trend scenario is able of resembling the observed data series. The area of traditional irrigated lands decreases to values around 7500 ha in response to the increase in both the real population but also the potential new population according to the successive land plans, confirming that this is effectively the key driving force inducing the accelerated loss of huerta in Murcia (Fig. 8.8).

The model ability to replicate the basic behaviour pattern of the system in the period 1995–2010 with no further model calibration against observed data, may be considered a sort of model validation with an independent dataset. This constitutes a valuable and rather uncommon feature regarding models of socio-ecological systems.

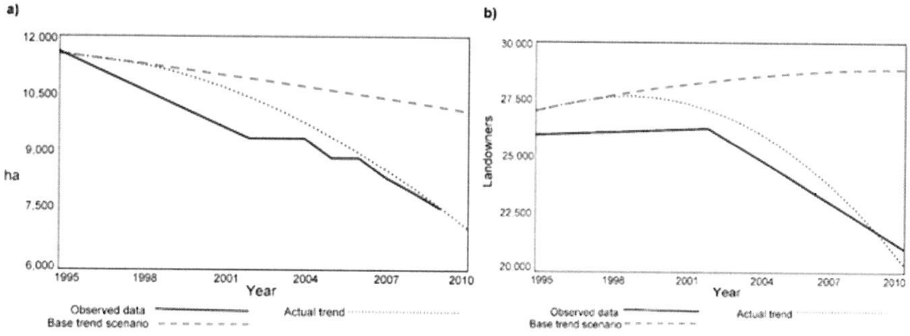

Fig. 8.8 Observed data and simulation results under the base tren scenario and the actual trend simulation for (**a**) the Huerta agroecosystem and (**b**) the number of landowners

In synthesis, obtained results show that the traditional irrigated land around Murcia is being lost at an accelerated rate. This will impact on the urban resilience of Murcia to climate change, regarding the capacity of the huerta to moderate the heat island effect of city. As a consequence, Murcia will be more vulnerable to the heat waves, which are expected to increase in intensity and frequency under the ongoing climate change. In order to improve the resilience of the city to the climate change impacts, the Huerta model will be applied to explore different land use and agricultural policy measures and their effectiveness in maintaining and recovering the Huerta periurban agro-ecosystem.

8.4 Vulnerability and Policy Assessment

8.4.1 Policies, Scenarios and Vulnerability

Dynamic simulation models are particularly useful as prospective tools (as opposed to predictive or forecasting tools). These kind of models capture the general structure and behaviour of models and are intended to explain the general dynamics of the system in the long run, not precise details in the short term. In coherence with this, dynamic models are not developed to tell what will happen, but to answer "what-if" questions. They are particularly useful at the level of strategic decisions and planning, more than at the managerial or operational level.

In the case of socio-ecological systems, including urban areas, these "what-if" questions can refer to actions that can be taken within the modelled system or to changes in the boundary conditions, this is, in the factors that condition the system but whose behaviour cannot be determined or decided within the system. These two different situations are usually distinguished with the terms *policies* (actions which can be implemented within the system to achieve certain objectives) and *scenarios* (external changes in the boundary conditions). Dynamic models can be applied to assess the expected effects of different policy measures and to explore the vulnerability of the system to certain scenarios, such as climate change or an economic recession.

Regarding resilience, there is a remarkable need of dealing with long-term dynamics, since resilient policies can only be successful if they consider long time horizons. Long-term planning is especially important when short-term decisions have long-term consequences, since it makes it possible to visualise key issues that may otherwise be missed. For this purpose, scenario development and policy assessment is one of the major tools to compare the potential outcomes of a variety of alternatives and to anticipate the long-term consequences of scenarios, policy decisions and actions (Banos-González et al., 2016).

In relation to policy measures, dynamic simulation models can be applied to (i) analyse the measures proposed by different agents and action plans; (ii) quantify their effects in terms of resilience indicators and thresholds; (iii) identify side-effects and trade-offs among objectives; (iv) determine the degree of uncertainty of the simulation results and (v) prioritize among measures.

The role of scenarios and policy assessment is illustrated with the water and energy nexus. Urban areas are highly dependent on other socio-ecological systems for the provision of resources, as water and energy. This dependence makes urban areas potentially vulnerable to possible shortages of such resources. The increasing proportion of urban population is emphasising this potential vulnerability. Therefore, it is important to anticipate the effects of different management options regarding resilience, as well as the urban vulnerability under different socio-economic and environmental scenarios, such as climate change. In the following section it is shown the combined use of dynamic simulation models, resilience indicators and thresholds to explore policies and scenarios regarding the water and energy nexus in the case of Fuerteventura.

8.4.2 Assessing Policy Measures for Urban Resilience in Fuerteventura

Energy and water are firmly interconnected and interdependent. Understanding this connection of water and energy has created a strong interest in exploring what Madani and Khatami (2015) named "the water-energy nexus". Water is used in energy mining and production, running turbines, cooling power plants, construction and operation of energy generation facilities and disposing their waste products. On the other hand, we need energy to purify, desalinate and transfer water (Hadian & Madani, 2013). This reciprocal dependency of water and energy is the core idea of the water-energy binomial.

Despite its high interdependence, managers of water and energy resources have conventionally operated independently (Madani & Khatami, 2015). There is a serious need for balancing the trade-offs between the different aspects of water-energy nexus management. A successful balance requires a good understanding, assessment and communication of the possible effects of different policy options and to share visions among policy makers, stakeholders and other agents, based on sound scientific knowledge.

This idea is even more challenging in the case of arid islands, such as Fuerteventura (The Canary Islands). In Fuerteventura, declared as Biosphere Reserve in 2009, there is a potential tradeo-off in terms of urban resilience objectives: on one hand, the marine water desalination increases the island resilience towards droughts and to an unexpected reduction in available resources in Fuerteventura; on the other hand, the dependence of a basic need as urban water supply from marine water desalination can be considered a vulnerability to the energy supply system, particularly if allochthonous, non-renewable energy sources are mainly used. In order to gain insights on the water and energy nexus, it has been applied the Fuerteventura Sustainability Model (FSM, Fig. 8.9), developed following the system dynamics methodology. It has been calibrated for the 1996–2011 period (Further details can be found in Banos-González et al., 2015). The FSM integrates five sectors: land use changes, socio-tourist sector, environmental quality, biodiversity and water resources. The model testing results (Banos-González et al., 2015) offer an adequate degree of model confidence to use it as a tool to analyse the main sustainability and resilience issues in Fuerteventura.

In Fuerteventura, urban water supply, both for resident (Fig. 8.10a) and tourist population, represents around 69% of the total water demand (around 12.5 Hm^3 in 2011). The net consumption per resident was 180 l per person and day, whereas tourists consumed around 378 l and 221 l per person and day in hotels and non-hotel tourist accommodations, respectively.

Surface water and groundwater pumping in Fuerteventura are clearly insufficient to fit the total water demands for the tourist and resident population, covering less

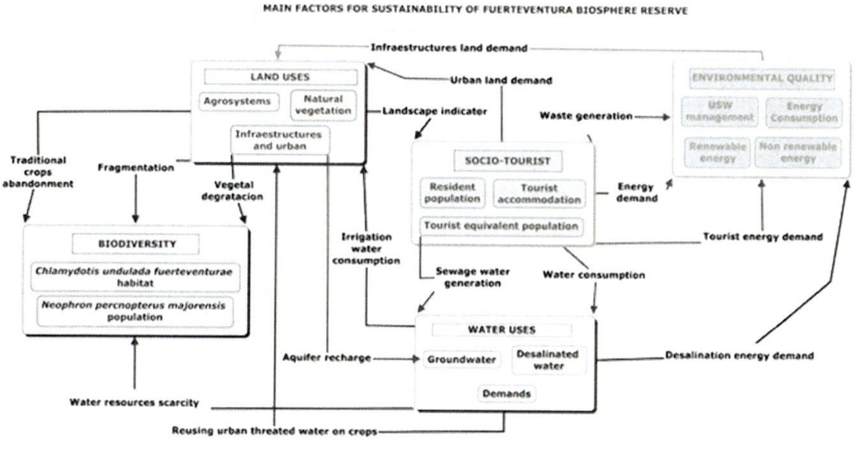

Fig. 8.9 Overview of the Fuerteventura Sustainability Model, FSM. (Further details can be found in Banos-González et al., 2015)

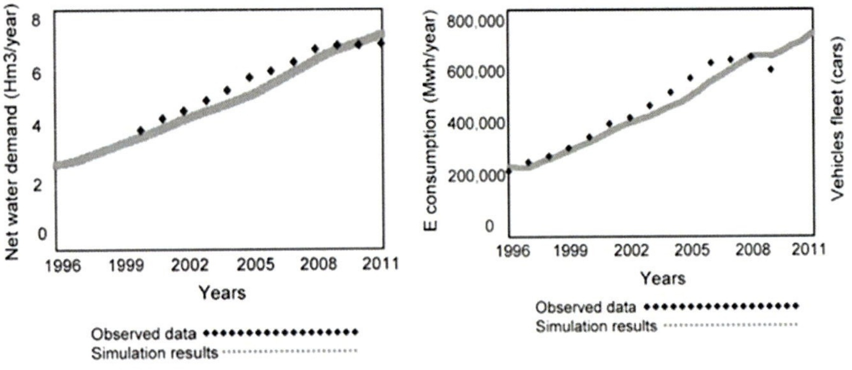

Fig. 8.10 Observed data in Fuerteventura and model simulation results for (**a**) Net water demand for water supply of resident population and (**b**) Total electric energy consumption

than 20%, in average. Therefore, seawater desalination is required to satisfy the remaining 80% of urban water demand. Moreover, desalination doubled along the 1996–2011 period. Although water desalination represents a minor component of total electricity demand (around 8.6% in 2011), such increase in water desalination also contributed to the rise in the total electric energy consumption (Fig. 8.10b). The share of renewable energy is around 2% of total energy use along the 1996–2011 period, meaning that practically all energy demands are fitted with external, non-renewable resources.

The FSM model has been applied to assess the objectives and policy measures contained in the Fuerteventura Biosphere Reserve Action Plan (AP, 2013), including those involving the water-energy nexus. This Plan includes the measure "100% renewable water", whose aim is to cover 100% of the electricity demand required for the supply of desalinated water with renewable energy in 2025, one of the basic guidelines of the AP. The FSM model, along with the integrated indicators and their thresholds, has allowed the assessment of the effectiveness of this policy measure (100% renewable water), as compared with the Business As Usual (BAU) run, under which no increase in renewable energy is implemented.

Three sustainability indicators related to energy and emissions of the Fuerteventura Biosphere Reserve Action Plan (Table 8.2) were selected and integrated into the FSM. A reference value or threshold was defined by each one of these three indicators.

Table 8.3 shows the simulation results for the three considered indicators under the policy measure of 100% Renewable Water, as well as under the Business As Usual (BAU) simulation for the 2012–2025 period. The estimated uncertainty for the expected results is also expressed, as 95% confidence intervals around the mean value of simulations, according to the Monte Carlo analysis.

The results under the Business As Usual simulation show that the urban demand for desalinated water would double between 2012 and 2025 due to the increase in population (both resident and tourist-equivalent population). The same increase

Table 8.2 Sustainability indicators on energy and emissions. Units and thresholds are also specified

Indicators	Units	Threshold	Meaning of the threshold	References of the thresholds
Per capita primary energy consumption (*PEpc*)	GJ/year*pc	<42	Minimum energy use required to reach a Human Development Index of at least 0.8, recommended by United Nation Development Programme.	Johansson and Goldemberg (2004)
Share of renewable energy (*SER*)	Dimensionless	>0.2	Renewable energy to represent at least 20% of total energy use in 2020 and 27% in 2030.	EC (2008 2015)
Per capita CO_2 emissions (CO_2pc)	metric tons CO_2/year*pc	<9.52	A 20% reduction in the per capita CO_2 emissions from 1990 levels. Based on 1999 value.	EC (2008)

Table 8.3 Simulation results in 2025 for the considered indicators under BAU and under the 100RW policy measure

Indicator	PEpc (GJ/ year*pc)	SER(dimensionless)	CO_2pc (metric tones CO2/ year*pc)
Threshold	<42	>0.2	<9.52
BAU	281.97±42.07	0.011±0.004	17.04±4.21
100RW	275.73±2.17	0.017±0.003	16.89±4.21

BAU Business As Usual, *100RW* Policy measure: 100% Renewable Water

would be expected for the electrical energy consumption. The total primary energy consumption would also increase, by around 52%. If all the power demand of desalination processes would be provided by renewable power, the SER indicator would increase by around 54%. However, the three indicators would exceed their thresholds under both simulations, even though the policy measure 100RW explicitly addresses the improvement of such indicators.

Whereas the seawater desalination, the main source of water on the island, has enabled to overcome the limitations of water scarcity on the socioeconomic activities, its negative side – a high energy consumption, an increased energy dependence and greenhouse and brine emissions – must be addressed (Meerganz von Medeazza & Moreau, 2007; Lattemann & Höpner, 2008; Melián-Martel et al., 2013), particularly in an island system with a low SER indicator. This dependence on allochthonous, non-renewable energy resources in Fuerteventura has not improved in the period 1996–2011, which represents a clear sign of unsustainability. Even more, the strong dependency of water availability for urban supply on energy consumption – more than the 80% of urban and tourist population water demand is covered by seawater desalination, – implies a high vulnerability of the whole socio-ecological system, even for basic needs, to socioeconomic changes such as those in the energy policies and markets, and to the ongoing global change (Kruyt et al., 2009). Climate change is expected to exacerbate the situation by increasing energy and water demands while at the same time available water resources are expected to decrease in Fuerteventura.

In synthesis, the combined use of the Fuerteventura dynamic model, along with the indicators and thresholds, allow to assess the relative effectiveness of planned policy measures, showing in this case that regarding the water-energy nexus such measures are far from achieving the proposed objectives. This type of assessment is essential to reorientate and better focus the actions aiming at substantially improving the urban resilience in Fuerteventura.

8.5 Uncertainty Assessment

8.5.1 Uncertainty and Urban Resilience

The assessment of socio-ecological system generally suffers from high levels of uncertainty. Complex models with many interactions among individual sources of uncertainty can increase the overall model uncertainty (Perz et al., 2013). Therefore, there is a need to identify potential sources of uncertainty and to quantify their impact on model outputs and on the application of each considered policy option. However, the existing models are often deterministic (Holzkämper et al., 2015; Uusitalo et al., 2015) highlighted that models which include the uncertainties related to management options may be of considerable added value for the decision makers. In fact, one of the properties of resilient systems is their ability to accept and cope with the inherent and ever-increasing uncertainty and change in today's world.

Regarding socio-ecological modelling, there are two different types of uncertainty. The first type arises from the lack of knowledge about the precise value of certain parameters and variables (for example the per capita net water consumption in certain urban typologies). This type of uncertainty can be reduced by specific studies aiming at improving the available information and data about such parameters and variables. This is particularly necessary in the case of parameters having a strong influence in model outcomes (high sensitivity parameters). The second type of uncertainty derives from intrinsic, non-reductible sources of variability in the socio-ecological system (for example the annual rainfall in Mediterranean environments). These sources of uncertainty should be explicitly accounted for in socio-ecological modelling, by determining appropriate uncertainty ranges to each parameter. Moreover, the interactions among variables can minimise or exacerbate the model response to all combined sources of uncertainty.

Among the methods across literature that enable to cope with uncertainty the Sensitivity Analysis (SA) is highlighted. SA is broadly used identify the key input variables and parameters that control model outputs (Schouten et al., 2014). An overview of SA methodologies can be found in Saltelli et al. (2005), Cariboni et al. (2007) and Refsgaard et al. (2007). The sensitivity analysis allows: (i) On one side, a detailed assessment of model robustness and, therefore, of the reliability of model outputs. (ii) On the other side, the quantification of the specific uncertainty associated to each model outcome under the considered scenarios and policies. The SA allows to answer questions as the following: How robust the conclusions derived from the model are?, How does uncertainty affect the assessment of policies and the vulnerability to certain external changes?

Different sensitivity analysis techniques can be applied, more specifically local sensitivity analysis or "One factor at a time" (OAT) and general sensitivity techniques by means of Monte Carlo simulation. OAT techniques allow us to determine the model sensitivity to each individual parameter. The sensitivity index ($S_{i,j}$, Jørgensen & Fath, 2011) is calculated as follows:

$$S_{i,j} = \left(\frac{OM_{i,t} - Om_{i,t}}{Ob_{i,t}} \right) / \left(\frac{PM_j - Pm_j}{Pb_j} \right) * 100$$

Where $S_{i,j}$ represents the sensitivity index of the target variable i to the parameter j; $OM_{i,t}$ and $Om_{i,t}$ are the maximum and minimum values of the *ith* target variable at time t; $Ob_{i,t}$ represents the base (default) model value of the *ith* target variable at time t; PM_j and Pm_j represent the maximum and minimum values of the *jth* parameter, respectively; and Pbj is the base model value of the *jth* parameter. The sensitivity index allows to discriminate between parameters with low ($Si,j < 10\%$), moderate ($10\% \leq Si,j < 50\%$), high ($50\% \leq Si,j < 100\%$) and very high sensitivity ($Si,j \geq 100\%$).

The general sensitivity analysis by means of Monte Carlo simulation (MC) allows us to assess the effects of a simultaneous variation of all sensitive parameters for each target variable. The variation coefficient (VC_i) of the target model variables shown by the Monte Carlo simulation is calculated as follows:

$$VC_{i,t} = \left(\frac{OM95_{i,t} - Om95_{i,t}}{\bar{O}_i} \right) * 100$$

Where VC_i represents the relative variation of the target variable i respect to its mean value using 95% confidence bounds; $OM95_i$ and $Om95_i$ are the maximum and minimum values of the ith target variable at time t using 95% confidence bound, and $\bar{O}_i$ is the mean value of the target variable i. According to the variation coefficient, the target model variables can show a low ($VCi < 50\%$), moderate ($50\% \leq VCi < 100\%$) and high response ($VCi \geq 100\%$) to changes in their respective most sensitive parameters.

The next section illustrates the uncertainty analysis using the FSM model in the Fuerteventura island.

8.5.2 *Assessing Uncertainty under Policy Measures and Scenarios in Fuerteventura*

As presented in the previous section, the FSM model was applied to assess different policy measures and scenarios in the Fuerteventura island (Banos-González et al., 2016). An extensive sensitivity analysis was also carried out to explore how uncertainty affects the model outcomes. One of the key issues in Fuerteventura is the rapid tourist development shown during the last decades. This Biosphere Reserve faces important challenges regarding its rural and urban resilience and the overall vulnerability of the island to this external driver. Using the FSM model, the embedded indicators and their thresholds, it has been assessed the Business As Usual simulation for the period 1996–2025. Figure 8.11 shows the results of the Montecarlo simulation in the period 1996–2025 under BAU for three key variables: the built-up urban area, the resident population and the equivalent tourist population.

The model includes three important indicators regarding the socio-tourist development: the ratio of tourist to resident population, the ratio between the tourist accommodation and resident population and the artificial land proportion (Table 8.4). Results show that the expected value of the ratio of tourists to residents would exceed the threshold value, whereas the two other indicators would not. However, when uncertainty is taken into account, it is clear that another indicator, the ratio between tourist accommodation and resident population, might also exceed the threshold.

In Fuerteventura this has been found not only in the case of the BAU simulation but also with different policy options and scenarios, as climate change, which have shown to be riskier than expected when only mean values are considered, since, when uncertainty is considered, such policies and scenarios present higher number of indicators exceeding their thresholds.

This highlights the importance of considering uncertainty in the decision process, when assessing different policy options. Some authors state that the precautionary principle should not represent a brake to decision-making, since inaction could have costly and unforeseeable impacts (Gee & Krayer von Krauss, 2005; Van der Sluijs, 2007). Uncertainty should be considered a normal component of decisions and, instead of inaction, it should appeal to the prudence of policy makers. In this sense,

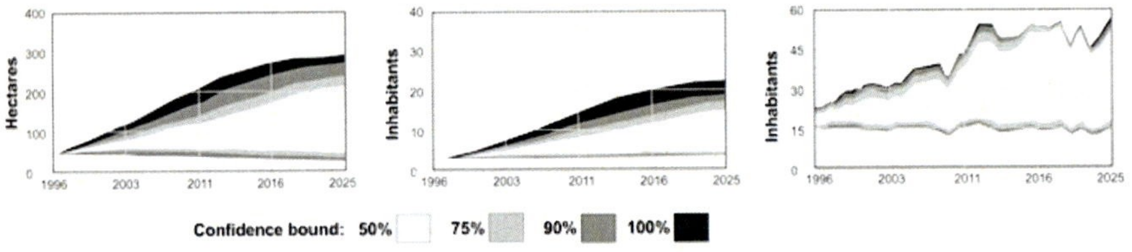

Fig. 8.11 Monte Carlo sensitivity analysis to changes in sensitive parameter values (local sensitivity over 50%) for the target model variables (left to right): Built-up urban area, Resident population and Equivalent tourist population

Table 8.4 Results of indicators related to the tourist development in Fuerteventura under the Business as Usual (BAU) simulation in the period 1996–2025

Indicator	Units	Threshold	Meaning of threshold	References for thresholds	BAU expected value and 95% confidence bound
Ratio of tourists to residents	Dimensionless	<0.3152	The ratio of tourist to local inhabitants should be lower than the threshold	Government of Canary Islands (2010)	0.329 ± 0.277 (0.053–0.606)
Ratio between tourist accommodation and resident population	Tourist beds/ inhabitant	<0.97	Ratio between tourist accommodation and resident population	Government of Canary Islands (2010)	0.618 ± 0.643 (0–1.261)
Artificial land proportion	%	<20	Proportion of area occupied by agriculture, urban use and infrastructures	Graymore et al. (2010)	6.83 ± 4.74 (2.09–11.57)

Units, threshold, references of thresholds, expected value and 95% confidence bounds are indicated

the precautionary principle should be applied taking into account the uncertainty analysis: the higher the uncertainty, the less risky the policy should be.

8.6 Conclusions. A Framework to Support Decisions for an Improved Resilience

In this chapter it has been presented an overall methodological framework to assess resilience in urban and other types of socio-ecological systems with the combined use of indicators, thresholds, dynamic simulation models, the assessment of policies and scenarios and the effects of uncertainty on model outcomes. As opposed to conventional Decision Support Systems (DSS), frequently based on assigning weights to each criteria, which are then combined into mixed indexes to find an optimal solution, here a different approach is proposed.

This approach allows to explicitly address the complex nature of real problems and the interactions and trade-offs among variables, by means of the indicators embedded in the model. The establishment of threshold values for each indicator provides a way to identify those policies that would exceed the concerned resilience thresholds. Following the rule "Threshold out, Measure out", any policy exceeding a threshold should be rejected or assigned the lowest priority. This approach avoids the use of indexes mixing non-reducible dimensions and, instead, keeps track of the positive and negative effects of each policy or scenario on the different environmental, economic and social factors involved in the concerned socio-ecological system. Finally, the participation of involved actors, particularly in the first stage (definition of objectives and indicators) and the final stage (assessment of policies and scenarios) is essential to ensure that the process effectively influences the decision process.

References

Adriaanse, A. (1993). *Environmental policy performance indicators*. Nature indicators survey. Ministry of Environment and Energy.

AP. (2013). *Action plan of the Fuerteventura biosphere reserve*. http://gestion.cabildofuer.es/fuerteventurabiosfera/

Banos-González, I., Martínez-Fernández, J., & Esteve-Selma, M. A. (2015). Dynamic integration of sustainability indicators in insular socio-ecological systems. *Ecological Modelling, 306*, 130–144.

Banos-González, I., Martínez-Fernández, J., & Esteve-Selma, M. A. (2016). Using dynamic sustainability indicators to assess environmental policy measures in Biosphere Reserves. *Ecological Indicators, 67*, 565–576.

Barlas, Y. (1996). Formal aspects of model validity and validation in system dynamics. *System Dynamics Review, 12*(3), 183–210.

Bell, S., & Morse, S. (2008). *Sustainability indicators. Measuring the immeasurable?* Earthscan.

Cariboni, J., Gatelli, D., Liska, R., & Saltelli, A. (2007). The role of sensitivity analysis in ecological modelling. *Ecological Modelling, 203*(1), 167–182. https://doi.org/10.1016/j.ecolmodel.2005.10.045

Comisión de las Comunidades Europeas. (1992). Programa Comunitario de Política y Actuación en materia de Medio Ambiente y Desarrollo Sostenible. COM (92) 23 FINAL. Instituto de Investigaciones Ecológicas.

EC. (2008). European Commission. Communication from the Commission-energy efficiency: Delivering the 20% target. COM 2008, 772 final.

EC. (2015). European Commission, Climate action, EU action 2030. Available from: http://ec.europa.eu/clima/policies/2030/index_en.htm

Figge, F., & Hahn, T. (2004). Sustainable value added—Measuring corporate contributions to sustainability beyond eco-efficiency. *Ecological Economics, 48*(2), 173–187. https://doi.org/10.1016/j.ecolecon.2003.08.005

Filatova, T., & Polhill, G. (2012). Shocks in coupled socio-ecological Systems: What are they and how can we model them? In *International Environmental Modelling and Software Society (iEMSs). International congress on environmental modelling and software managing resources of a limited planet, Sixth Biennial meeting*, Leipzig, Germany.

Gee, D., & Krayer von Krauss, M. P. (2005). Late lessons from early warnings: Towards precaution and realism in research and policy. *Water Science & Technology, 52*(6), 25–34.

Gonçalves, C., & Marques da Costa, E. (2013). Framework and indicators to measure urban resilience. Communication at: AESOP-ACSP Joint Congress . 15–19 July 2013. Dublin.

Government of Canary Islands. (2010). Presidencia de Gobierno. Viceconsejería de Turismo. Observatorio turístico: Estadísticas y Estudios.

Granda Leon, M., González Camba, S., & Calvopiña Carvajal, V. (2013). Medición de pobreza en Galápagos. Informe Galápagos 2011–2012. DPNG, CGREG, FCD y GC. Puerto Ayora, Galápagos, Ecuador.

Graymore, M. L., Sipe, N. G., & Rickson, R. E. (2010). Sustaining human carrying capacity: A tool for regional sustainability assessment. *Ecological Econòmics, 69*(3), 459–468. https://doi.org/10.1016/j.ecolecon.2009.08.016

Guyot-Tephany, J. et al. (2013). Usos, percepciones y manejo del agua en Galápagos. pp. 67–75. Informe Galápagos 2011–2012. DPNG, CGREG, FCD y GC. Puerto Ayora, Galápagos, Ecuador.

Haag, D., & Kaupenjohann, M. (2001). Parameters, prediction, post-normal science and the precautionary principle – A roadmap for modelling for decision-making. *Ecological Modelling, 144*, 45–60.

Hadian, S., & Madani, K. (2013). The water demand of energy: Implications for sustainable energy policy development. *Sustainability, 5*(11), 4674–4687.

Hanley, N., McGregor, P. G., Swales, J. K., & Turner, K. (2009). Do increases in energy efficiency improve environmental quality and sustainability? *Ecological Economics, 68*(3), 692–709. https://doi.org/10.1016/j.ecolecon.2008.06.004

Hodbod, J., & Adger, W. N. (2014). Integrating social-ecological dynamics and resilience into energy systems research. *Energy Research & Social Science, 1*, 226–231. https://doi.org/10.1016/j.erss.2014.03.001

Holzkämper, A., Klein, T., Seppelt, R., & Fuhrer, J. (2015). Assessing the propagation of uncertainties in multi-objective optimization for agro-ecosystem adaptation to climate change. *Environmental Modelling & Software, 66*, 27–35. https://doi.org/10.1016/j.envsoft.2014.12.012

Jakeman, A. J., & Letcher, R. A. (2003). Integrated assessment and modelling: Features, principles and examples for catchment management. *Environmental Modelling & Software, 18*(6), 491–501.

Johansson, T. B., & Goldemberg, J. (2004). Energy and the challenge of sustainability overview: 2004 update, United Nations Development Programme, United Nations Department of Economic and Social Affairs, and World Energy Council. Available from: http://www.undp.org/seed/eap/activities/wea/

Jørgensen, S. E., & Fath, B. (2011). *Fundamentals of ecological modelling*. Elsevier.

Kampmann, C. E., & Oliva, R. (2008). Structural dominance analysis and theory building in system dynamics. *Systems Research and Behavioral Science, 25*(4), 505–519. https://doi.org/10.1002/sres.909

Kelly, R. A., Jakeman, A. J., Barreteau, O., Borsuk, M. E., ElSawah, S., Hamilton, S. H., & Voinov, A. A. (2013). Selecting among five common modelling approaches for integrated environmental assessment and management. *Environmental Modelling & Software, 47*, 159–181. https://doi.org/10.1016/j.envsoft.2013.05.005

Khazai, B., Bendimeard, F., Cardona, O., Carreño, M. L., Barbat, A., & Burton, C. (2015). A guide to measuring urban risk resilience. Principles, tools and practice of urban indicators. *Earthquakes and Megacities Initiative (EMI), The Philippines*.

Kruyt, B., van Vuuren, D. V., De Vries, H. J. M., & Groenenberg, H. (2009). Indicators for energy security. *Energy Policy, 37*(6), 2166–2181. https://doi.org/10.1016/j.enpol.2009.02.006

Lacitignola, D., Petrosillo, I., Cataldi, M., & Zurlini, G. (2007). Modelling socio-ecological tourism-based systems for sustainability. *Ecological Modelling, 206*(1), 191–204. https://doi.org/10.1016/j.ecolmodel.2007.03.034

Lancker, E., & Nijkamp, P. (2000). A policy scenario analysis of sustainable agricultural development options: A case study for Nepal. *Impact Assessment and Project Appraisal, 18*(2), 111–124. https://doi.org/10.3152/147154600781767493

Lattemann, S., & Höpner, T. (2008). Environmental impact and impact assessment of seawater desalination. *Desalination, 220*(1), 1–15. https://doi.org/10.1016/j.desal.2007.03.009

Li, F. J., Dong, S. C., & Li, F. (2012). A system dynamics model for analyzing the ecoagriculture system with policy recommendations. *Ecological Modelling, 227*, 34–45. https://doi.org/10.1016/j.ecolmodel.2011.12.005

Liu, G., Yang, Z., Chen, B., & Ulgiati, S. (2014). Emergy-based dynamic mechanisms of urban development, resource consumption and environmental impacts. *Ecological Modelling, 271*, 90–102. https://doi.org/10.1016/j.ecolmodel.2013.08.014

Madani, K., & Khatami, S. (2015). Water for energy: Inconsistent assessment standards and inability to judge properly. *Current Sustainable/Renewable Energy Reports, 2*(1), 10–16. https://doi.org/10.1007/s40518-014-0022-5

Martínez-Fernández, J., Esteve-Selma, M. A., Baños-González, I., Carreño, F., & Moreno, A. (2013). Sustainability of Mediterranean irrigated agro-landscapes. *Ecological Modelling, 248*, 11–19. https://doi.org/10.1016/j.ecolmodel.2012.09.018

Martínez-Fernández, J., Banos-Gonzalez, I., Esteve-Selma, M. A., & Carrión-Tacuri, J. (2016). *IDIS: una metodología de definición de indicadores de sostenibilidad del agua. Aplicación al caso de las Islas Galápagos*. Proceedings of the IX Congreso Ibérico de Planificación y Gestión del Agua. Agua, ciudad y salud de los ecosistemas. J. Raúl Navarro García y Laura Sánchez Gallardo (Eds.). Fundación Nueva Cultura del Agua.

Martínez-Moyano, I. J., & Richardson, G. P. (2013). Best practices in system dynamics modeling. *System Dynamics Review, 29*(2), 102–123. https://doi.org/10.1002/sdr.1495

Meerganz von Medeazza, G., & Moreau, V. (2007). Modelling of water–energy systems. The case of desalination. *Energy, 32*(6), 1024–1031. https://doi.org/10.1016/j.energy.2006.10.006

Melián-Martel, N., Sadhwani Alonso, J. J., & Pérez Báez, S. O. (2013). Reuse and management of brine in sustainable SWRO desalination plants. *Desalination and Water Treatment, 51*(1-3), 560–566. https://doi.org/10.1080/19443994.2012.713567

Moldan, B., Janoušková, S., & Hák, T. (2012). How to understand and measure environmental sustainability: Indicators and targets. *Ecological Indicators, 17*, 4–13. https://doi.org/10.1016/j.ecolind.2011.04.033

Mori, K., & Christodoulou, A. (2012). Review of sustainability indices and indicators: Towards a new City Sustainability Index (CSI). *Environmental Impact Assessment Review, 32*(1), 94–106.

Perz, S. G., Muñoz-Carpena, R., Kiker, G., & Holt, R. D. (2013). Evaluating ecological resilience with global sensitivity and uncertainty analysis. *Ecological Modelling, 263*, 174–186. https://doi.org/10.1016/j.ecolmodel.2013.04.024

Proelss, A., & Houghton, K. (2012). The EU Common Fisheries Policy in light of the precautionary principle. *Ocean & Coastal Management, 70*, 22–30. https://doi.org/10.1016/j.ocecoaman.2012.05.015

Reed, M. S., Fraser, E. D., & Dougill, A. J. (2006). An adaptive learning process for developing and applying sustainability indicators with local communities. *Ecological Economics, 59*(4), 406–418. https://doi.org/10.1016/j.ecolecon.2005.11.008

Refsgaard, J. C., van der Sluijs, J. P., Højberg, A. L., & Vanrolleghem, P. A. (2007). Uncertainty in the environmental modelling process–A framework and guidance. *Environmental Modelling & Software, 22*(11), 1543–1556. https://doi.org/10.1016/j.envsoft.2007.02.004

Rodríguez-Rodríguez, D., & Martínez-Vega, J. (2012). Proposal of a system for the integrated and comparative assessment of protected areas. *Ecological Indicators, 23*, 566–572. https://doi.org/10.1016/j.ecolind.2012.05.009

Saltelli, A., Ratto, M., Tarantola, S., & Campolongo, F. (2005). Sensitivity analysis for chemical models. *Chemical Reviews, 105*(7), 2811–2828.

Schouten, M., Verwaart, T., & Heijman, W. (2014). Comparing two sensitivity analysis approaches for two scenarios with a spatially explicit rural agent-based model. *Environmental Modelling & Software, 54*, 196–210. https://doi.org/10.1016/j.envsoft.2014.01.003

Suárez, M., Gómez-Baggethun, E., Benayas, J., & Tilbury, D. (2016). Towards an urban resilience index: A case study in 50 Spanish cities. *Sustainability, 8*, 774. https://doi.org/10.3390/su8080774

Uusitalo, L., Lehikoinen, A., Helle, I., & Myrberg, K. (2015). An overview of methods to evaluate uncertainty of deterministic models in decision support. *Environmental Modelling & Software, 63*, 24–31. https://doi.org/10.1016/j.envsoft.2014.09.017

Van der Sluijs, J. P. (2007). Uncertainty and precaution in environmental management: Insights from the UPEM conference. *Environmental Modelling & Software, 22*(5), 590–598. https://doi.org/10.1016/j.envsoft.2005.12.020

Vidal-Legaz, B., Martínez-Fernández, J., Sánchez Picón, A., & Pugnaire, F. (2013). Trade-offs between maintenance of ecosystem services and socio-economic development in rural mountainous communities in southern Spain: A dynamic simulation approach. *Journal of Environmental Management, 131*, 280–297. https://doi.org/10.1016/j.jenvman.2013.09.036

Voinov, A., & Shugart, H. H. (2013). 'Integronsters', integral and integrated modeling. *Environmental Modelling & Software, 39*, 149–158. https://doi.org/10.1016/j.envsoft.2012.05.014

Chapter 9
Sustainable and Resilience Descriptors for the Xochimilco-Tláhuac Lacustrine Area at México City

Enrique Mendieta-Márquez, Octavio F. González-Castillo, and Margarita Juárez-Nájera

Abstract Sustainability as well as resilience are dynamic, multidimensional concepts, subject to diverse interpretations; from a global perspective, they take into account different ways of life and local cultures, and require the development of capacities and the use of knowledge from different stakeholders, in order to build their social significance through interdisciplinary problem analysis. The objective of this work is to propose a set of action guides to drive the intervention of a conservation area (Xochimilco-Tláhuac polygon, XT-P), towards an idealized scenario of resilience and sustainability, while considering its relevant contribution as part of the region called 'Green Horseshoe', within Mexico City Metropolitan Area (MCMA). A set of descriptors, based upon a method named Plus Planning Cycle, originally designed by the International Centre for Sustainable Cities (ICSC, The sustainable cities: PLUS planning cycle of Calgary, Canada. Retrieved from http://www.icsc.ca/content/dmdocuments/plus_planning_cycle.pdf, 2008) for a study case in Calgary, is proposed. These descriptors were distributed among the five kinds of sustainable domains used by ISCC: natural, built, economic, social, and governance. To support the proposal, a summary of the principles on which the descriptors were built, as well as specific action guides for each domain, are included.

Keywords Sustainability · Resiliences · Urban socioecosystem · Lacustrine area · Action guides

E. Mendieta-Márquez (✉) · O. F. González-Castillo · M. Juárez-Nájera
Universidad Autónoma Metropolitana-Iztapalapa, Mexico City, Mexico
e-mail: enme@xanum.uam.mx; ogc@xanum.uam.mx; mjn@azc.uam.mx

O. F. González Castillo et al. (eds.), *Urban Resilience: Methodologies, Tools and Evaluation*, Resilient Cities, https://doi.org/10.1007/978-3-031-07586-5_9

9.1 Introduction

Urban processes in Mexico City have been increasing, particularly during the last 50 years. The city's third cycle of urban expansion has been mainly directed towards the southeastern part of the Valley of Mexico, reaching the conservation soil areas collectively known as the 'Green Horseshoe (GH)', which have represented until now its natural reserve. This expansion has degraded rural soil, particularly in the lake areas of *Xochimilco* and *Tláhuac* (XT), which were included in the UNESCO World Heritage List in 1987.

Human impact in the XT area has predominantly affected its water resources, due to the overexploitation of the aquifer and the drying of wells. Water shortage has caused changes in the traditional way in which local inhabitants have been growing vegetables and fruits (*chinampas*) which, in turn, has led to land use changes and to further urbanization increases, reducing the system's resilience capacity, and finally resulting in profound alterations in the life of people that inhabit this area.

In a similar way, environmental deterioration has also affected the biodiversity in the XT area; i.e., the loss of vegetative cover has led to soil erosion and biodiversity decrease, changes that are affecting, not only its sustainability and resilience capacity, but also that of the Green Horseshoe and of the Valley itself.

During 2007–2008, Mexico City's Ministry of Urban Development and Housing (SEDUVI) appointed a group of researchers from the Metropolitan Autonomous University (UAM) to carry out a study designed to evaluate future scenarios for an area called the *Xochimilco-Tláhuac* Polygon (XT-P), as part of a programme for the development of the so-called Sustainable Integral Metropolitan Zone (SIMZ). This Polygon represents an intervention area, where productive programmes will be designed and operated, in order to favour the sustainable development and the resilience of the zone.

The task force assembled for such study was divided in several groups, one of which was focused on defining in an interdisciplinary way, an operational resilience and sustainability vision for the XT-P. The Interdisciplinary Sustainability (IS) Team incorporated six UAM's researchers from three different campuses and five departments, all of them members of UAM's Institutional Plan towards Sustainability (PIHASU), which integrate a common and interdisciplinary view from the Natural, Social and Engineering sciences.

Initially, the IS Team compiled a preliminary document, based upon a group of descriptors that visualized the XT-P idealized state considering it, not independently, but a part of the GH. The IS Team also discussed some of the action guides needed to approach such condition, in an effort towards the construction of viable indicators that could later be used to evaluate the state of the system's sustainability and resilience throughout specific periods of time.

9.2 Mexico City Metropolitan Area (MCMA)

Mexico City Metropolitan Area (MCMA) has its seat in what was five centuries ago, a lacustrine area of 2000 km^2. MCMA emerges in a process of urbanisation from Mexico City centre towards its periphery, and through time, has absorbed villages, small towns, and rural areas from other federal states. It forms a set of contiguous administrative political units, socially and culturally integrated, whose complexity is reflected in the government structures that deal with its operation within their social, economic, environmental, and territorial processes. While only representing a thousandth part of the national territory, MCMA gathers one sixth of Mexico's population (approximately 22 million inhabitants); generates a third of its GDP; centralizes more than 60% of the banking activity and more than three quarters of financial savings; concentrates more than half of Mexican industry, and 75% of its cars; consumes 150 times more energy per unit area than the rest of the country, and annually releases more than 5 million tons of pollutants (UNEP, Centro GEO, 2004; Garza, 2000).

A combination of environmental and anthropogenic factors have significantly reduced life quality in MCMA, such as high population increase rates over the last 50 years; mobility strategies focused on cars; industrialization processes in the northern part of Mexico City under an economic model seeking development, but failing to consolidate industrial facilities with high levels of competitiveness and sustainability profiles; and poorly implemented urbanization plans, where almost all lakes have been drained, forest areas have been cut down, and farmland is buried under the urban soil. All of these factors contribute to the following problematic situation in the MCMA:

- Atmospheric pollution constantly exposes population to airborne pollutant concentrations above tolerable levels, and occasionally generates acid rain.
- Aquifers are overexploited, and 40% of water wells are polluted with faecal material, and in some cases with sulphides.
- All surface water bodies and rivers have irreversible damage.
- In order to preserve *Xochimilco* Lake as a permanent structure, it is being filled with treated water, but it also receives industrial and domestic wastewater.
- Wooded areas on lower slopes of the hills in the southern and western part of the city, as well as scrub forests and cloud forests have practically disappeared (Pisanty, 2000).
- Change of land use provokes deforestation at a rate of 239 ha/year and soil erosion, estimated in 1998, generated 834,000 m^3 of silt.
- Real estate developments, as well as legal and illegal land occupation, have exerted excessive pressure on ecosystems, leading to evident degradation of environmental services.
- Deterioration of public spaces and decrease in green areas.
- Excessive commute time and poor transport conditions.
- Water, food and energy dependence from other regions, as well as inadequate waste management.

All these elements have affected the resilience processes increasing vulnerability to the following risks:

- Flood.
- Seismicity.
- Vulcanism.
- Landslides in canyons and hillsides.
- Differential floor subsidence.
- Infectious and parasitic water-borne diseases.
- Atmospheric, water and soil pollution.

Therefore, different areas within the MCMA can be considered as environmentally impoverished or even exhausted, especially in its southwestern part, where threats to people's well-being are present (e.g., people living in gullies) (Ezcurra, 2000; Mazari & Noyola, 2000). Unfortunately, quality of life varies by regions and socio-economic strata, thus decreasing the possibilities to reduce risks and mitigate damages.

9.3 Xochimilco-Tláhuac Polygon (XT-P)

XT-P is a conservation area inside the MCMA, located in the southeastern part of the valley. It is a floodable area, which is a discharge area of groundwater flow. It is home to many species of aquatic and terrestrial flora and fauna, some of them vulnerable and with a very restricted geographical distribution. Therefore, the polygon captures water, controls climate, is a source of agroforestry, conserves biodiversity, and preserves regional ecological balance.

Physiographically, XT-P is located in the Neovolcanic province, *Anahuac* Lakes and Volcanoes Sub-province. The lake area is related to rivers and originated in the upper Tertiary and early Quaternary periods, belonging to the *Pánuco* River region, within the *Moctezuma* River watershed and the *Texcoco-Zumpango* subwatershed. Topographically, XT-P has a slope ranging from 0% to 5% and its geomorphology consists of lacustrine flat plains, riverine floodplains which mark the transition between the valley and the mountains, and the saline lake located on the border of *Xochimilco-Tláhuac* and *Iztapalapa* plains.

Its climate is considered temperate sub-humid, with summer rains. The average annual rainfall is 620.4 mm, mainly present between May and October. The average annual temperature ranges between 12 and 18 °C. The soils have a high content of organic matter, are dark, and influenced by the presence of a nearby aquifer, mainly composed of clay mixed with volcanic ash, with silt-clay textures.

Data for 2006 show that *Xochimilco* county had a total surface of 11,812.585 ha, with approximately 70% used for agriculture, 20% as urban area, and 10% of grassland and forest, while *Tláhuac* county had 8,591.209 ha distributed in a similar fashion, although its urban area is slightly larger, while agricultural use is comparatively diminished. Due to its recent transformation into an urban area and its particular topographical characteristics, XT-P has an intricate network of local roads, with few main avenues connecting it with the central part of MCMA. Mobility infrastructure includes bus and taxi routes for both counties, while only *Xochimilco* county has several rail commute stations.

Demographically XT-P had 219,279 inhabitants in 2000, a third of the total population of *Xochimilco* and *Tlahúac* counties, estimated at 668,000, and almost 4% of the Mexico City's total population for that year (8,605,239 inhabitants). Population between 1 and 14 years represented 30% of the total, while those aged 60 or more were only 7%, with a clear tendency to increase; 84% of *Tláhuac* and *Xochimilco* population is native. These numbers have remained essentially the same, at least until the last census (INEGI, 2015).

Spanish is the official language; however, *Náhuatl* (Aztec language) is the second spoken language. More than 2% of its population in 2000 spoke it, around 8000 individuals. XT-P is also a repository of a regional culture that is composed of basic religious elements related to agriculture and its rituals. The use of *trajineras* (boats driven by long wooden pilings) has been the means of transportation for regional trade. There are 189 km of interconnected channels within the XT-P.

Data for 2015 show that more than 97% of the population over 15 years old in both counties are literate, the same percentage in MCMA; 68% had some sort of post-primary education (6 years), more than half had upper secondary level instruction (high school), and between 15% and 20% completed higher education (university degree). Women who have reached high school show a preference for administrative and communication areas, while men are inclined to technological branches.

Related to health issues, beneficiaries of a medical institution reach 83% of the population for *Tláhuac* county and only 76.3% for *Xochimilco* county. The Mexican Social Security Institute has no high-level medical clinics in these areas, and the Institute of Security and Social Services for State Workers has a general practice medical clinic. The only third-level hospital in the region is located in *Tláhuac* county, and is managed by the Ministry of Health, focusing on paediatric care, maternity and emergencies.

For cultural issues, XT-P has a dozen public forums and twenty public libraries. In *Xochimilco* county there is only one museum run by the State, which is located in the town of *Santa Cruz Acalpixca* and houses a collection of objects related to the *Cuailama* ceremonial centre. Of great importance is the *Dolores Olmedo Patiño* Museum, located in *La Noria;* it houses numerous works of *Diego Rivera* and *Frida Kahlo*. Near this museum is located the *Carlos Pellicer* Forum, which serves as a theatrical stage.

Since 900 dc, people in XT-P developed a cultivation system called *chinampas*, false islands made with mud and silt, artificially anchored by poles, still in use today. Thus, XT-P stands out as an agricultural centre, producing corn, beans, peppers, pumpkins, tomatoes, *chilacayotes*, *chia*, *huauzontle*, and green beans, and since the colonial era, produced plants in large quantities for local and external consumption, such as turnips, leeks, broccoli, beets, celery, onions, carrots, lettuce, cabbage, and *quelites*. *Chinampas* provide basic food to people as well as a surplus to exchange or sell; however, with time the length of water channels has decreased, as well as the profitability of *chinampa* agriculture. Currently, the *chinampa* area, has almost entirely been given the status of area subject to ecological conservation, along with the irrigation districts found northwest of the XT-P.

The economically active population in *Tláhuac* and *Xochimilco* counties accounted for 38% of total inhabitants in 1986 and reached 54% in 2015. Population employed in the primary sector was less than 10%, and those employed within the tertiary sector accounted for almost 75% of the total.

In XT-P, almost 16% of the working population does not receive a fixed income or receive less than the minimum monthly salary, and 33% only between 1 and 2 minimum monthly wages of labour income. The rest of economically active population gets very low or no earned income, reflecting salaries less than 2000 mexican pesos monthly. At this point, an incentive programme for agricultural production, with its incorporation to cultural activities or alternative tourism, seems imperative.

Due to population growth, partly explained by the movement towards peripheral regions in search of available land, XT-P has recently faced strong pressure to convert productive land to residential use. Thus, despite of regulations and established land management programmes, invasions have proliferated and small housing construction has increased. Regarding the latter, in XT-P 78% of private homes have concrete slab ceilings and brick walls, while 88% of private households have tap water, drainage and electricity, numbers are lower than the average for MCMA.

Since the early nineteenth century, Mexico City water management system has continuously altered the *Xochimilco* and *Tláhuac* lake system. Excessive water extraction has caused a decrease in the aquifer's hydraulic load, which in turn has favoured the consolidation of soil and its collapse. In order to counter the phenomenon, treated wastewater is ministered, which has caused deterioration in water quality, affecting the structure and function of aquatic and terrestrial ecosystems.

All the aforementioned conditions lead us to consider that XT-P is subject to environmental, social and economic difficulties that affect the socioecosystem resilience, and endangers its maintenance in time. In order to prevent further deterioration, an ideal image of this region should be built, based upon a set of descriptors that can allow us to quantitatively monitor changes, and develop action guides to drive the XT-P towards resilient and sustainable development scenarios.

9.4 Method

During this research, the use of complex systems approach and an interdisciplinary work platform were privileged (Clayton & Radcliffe, 1996; Checkland, 1997; UAM, 2006), seeking to resignify the relationship between natural and human spheres as a basis for proposing and implementing viable management strategies that contribute to resilience and sustainability in urban socioecosystems. The following considerations were also taken into account:

- Pollution, species extinction, and poverty, are features of environmental degradation, representing anomalies in the current development model.
- Biological and cultural diversity have intrinsic value, and are key factors contributing to the stability of the life support systems in the planet.
- Technical solutions are necessary, but not sufficient to promote resilience and to approach sustainability. In order to achieve resilience and sustainability, a profound transformation in the way human beings have seeked satisfaction of their needs is required.
- Concepts of sustainability and development are dynamic mental constructs (abstract systems designed by man) that need to be discussed, agreed and continually reviewed locally.
- Resilience and sustainability are too complex for deductive analysis and not random enough to be statistically analysed. In this interface, the systems perspective offers a better methodological approach, as well as a multidimensional structure for information integration coming from different disciplines.
- Approaches to resilience and sustainability show their influence on the urban planning process management basically in four ways: (a) criticizing a purely reductionist perspective while reinforcing an expansionist perspective; (b) expanding the relevant areas and dimensions, as well as the spatial and temporal horizons considered; (c) expanding stakeholder responsibility and driving them towards a more participatory and committed design and; (d) promoting inter and transdisciplinary work.

An initial set of elements for building up prospective scenarios defining idealized futures for XT-P was integrated through group discussion, following a strength-weakness-opportunities-threats analysis (SWOT), based upon information gathered from previous UNESCO sponsored investigations in the area (López 2004; UNESCO et al., 2006). Using these ideas as a starting point, the IS Team devised a set of descriptors based upon the method named *Plus Planning Cycle*, originally developed by the *International Centre for Sustainable Cities* (ICSC). Descriptors were assigned to one of the five kinds of sustainable domain (natural, built, economic, social, and governance) used by the ICSC in a study case for the City of Calgary, Canada (ICSC, 2008). The IS Team also proposed action guides that could help the system approach its idealized state, based upon these descriptors. The method is schematically presented in Fig. 9.1.

9.5 Results

The IS Team proposed the following scenario as the most appropriate to guide XT-P toward its idealized state (resilient and sustainable) (SEDUVI-UAM, 2008):

- Proper water management is enforced; subsidence is stabilized and water quality is increased.
- *Chinampas* are still used, and they remain as an alternative for local production.

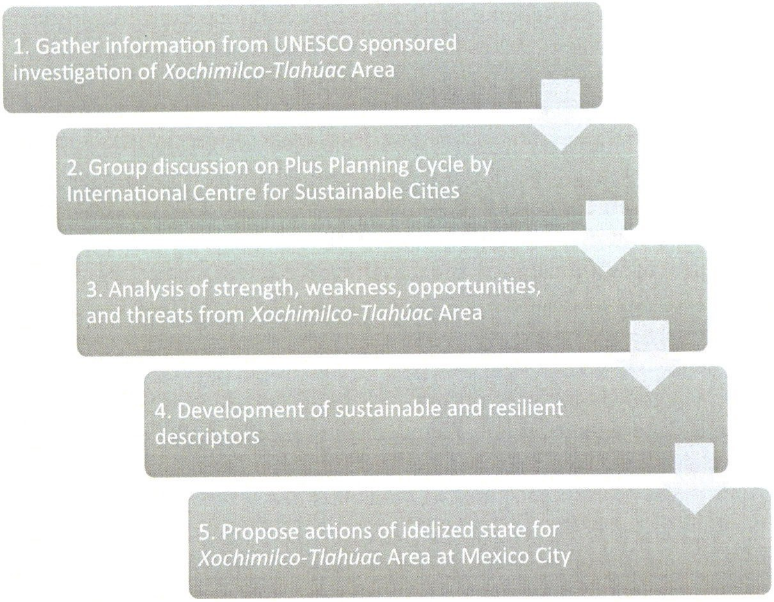

Fig. 9.1 Planning of the study in five stages

- Agricultural production is diversified (e.g. by using organic and flower cultures), has increased its commercial value, and depends mainly upon young people's work.
- XT-P is self-sufficient regarding the treatment of its liquid and solid wastes.
- New and dynamic social organizations deal with the identification, analysis and settlement of problems, by means of the use of efficient management strategies.
- Everyone considers communication processes, as well as people's information rights, equity, justice, tolerance, and diversity as key values.
- Traditional, cultural and natural values are preserved and spread.
- Sustainability and resilience assessment and monitoring are a common practice.
- High levels of governance are reached.

Based upon these general ideas, the following set of descriptors was proposed by the IS Team. They have been distributed between the different domains considered by the ICSC as follows:

5.1 Natural Setting: Water (6), Air (4), and Land (4)
5.2 Built Setting: Transportation (6), Urban Services (7), and Energy (5)
5.3 Economic Setting (5)
5.4 Social Setting (4)
5.5 Governance Setting (7)

9.5.1 *Natural Setting*

9.5.1.1 Water

Water available for conservation soil areas within the XT-P is currently been obtained from the *Cerro de la Estrella* treatment plant, but it is clearly insufficient for maintaining a hydraulic equilibrium within the system. Other strategies could also contribute to ensure its sustainable use (e.g. rain capture in mountain zones within the Green Horseshoe, reuse of treated water, controlled extraction from the aquifer). Non-drinkable water, provided by infiltration facilities *in situ,* as well as grey water treatment could also be alternatives for human activities not directly related to domestic uses.

Drinkable water distribution networks must be limited to those areas within the XT-P that are not considered as conservation soil, while autonomous or semi-autonomous systems should instead be used in areas regarded as such. Due to its environmental impacts, new human settlements must be legally prevented within conservation areas. Small treatment plants, mainly oriented toward self-consumption, requiring minimal amounts of energy expenditure and producing low-polluting emissions through the use of environment friendly technologies, could be built for households already located within these areas.

Because a good part of the XT-P can be considered as an agroecological system, Mexico City's Government, in coordination with all stakeholders, must define areas suitable for other uses, such as habitable, agricultural, cattle raising activities or ecotourism, taking into account factors such as hydraulic infrastructure, drainage systems, and conservation needs. Channel border reinforcement and silt removing programmes could prevent the drying of areas employed for agricultural production.

Similarly, complementary measures that could be useful to ensure water quality within the XT-P include: (a) avoiding the operation of garbage dumps and landfills; (b) fertilizing practices using chemicals must be reduced, while favouring the use of biologicals for this purpose, and; (c) efforts must be coordinated for encouraging the use of housekeeping biodegradable products. Assessment is imperative. Sustainability and resilience indicators must be designed for each different water use in order to seek efficiency in the management of hydric resources by all stakeholders, while reducing water losses during its distribution or use. Mechanisms for emergency situation responses must be designed.

WATER Action Guides to Drive XT-P Toward Resilient-Sustainable Scenario

1. Water must be obtained through sources and processes that allow hydric equilibrium throughout time.
2. Distribution systems must assure quality water for population while preventing leaks.
3. Different water qualities must be provided for non-drinkable uses, (e.g domestic, agricultural and recreational), while maintaining each of their current approved standards.
4. Sewage water disposing systems must be designed in such a way as to ensure their separation from water distribution networks, minimizing the possibility of becoming themselves pollution sources.
5. Technological packages that could increase water productivity for all users must be implemented.
6. Water treatment plants and other similar facilities must be built in order to supply water with quality features that can make it available for reuse.

9.5.1.2 Air

According to its status as a conservation area, there should be no significant emissions from fixed sources (e.g., industrial and service sector facilities) located within the XT-P. The presence of important avenues and freeways close to this area, such as *Periférico* and *División del Norte Avenue*, could have a substantial impact in the air quality of this zone, while the effect of wind-driven pollution coming from the northeastern part of the city also cannot be put aside. Therefore, the agricultural use of the land in an extensive part of the intervention area should be maintained in order for it to continue functioning as a 'purifying lung' for the SIMZ. However useful this activity can be, traditional agriculture based upon slash and burn techniques poses a threat to the area, if fires are not kept under tight control; extensive tree sets are present, so they can also be damaged. Because conflagrations are not unusual in the dry season, pollution resulting from ashes and coal particles can be a real problem for human and animal populations living in the area. Surveillance and fire control programmes should be implemented, particularly those focused upon wastelands close to housing areas or abandoned *chinampas*, and these activities must allow the active participation of people living in this area.

Particles derived from human activity are probably not the main reason for air-quality loss in the XT-P. Due to animal (bovine, ovine or canine) uncontrolled defecation, biologically originated pollution can also constitute an important health problem, which has not been currently assessed. Even humans can be a source of this kind of air pollution when living in areas lacking piped drainage systems. The construction of facilities adapted to the conservation category of the region should be encouraged; stool pickup and treatment programmes, as well as educational efforts must be done to minimise this problem.

AIR Action Guides to Drive XT-P Toward Resilient-Sustainable Scenario

1. Gas and particle emission-diminishing technologies must be supported, both for fixed and mobile sources, according to the approved normativity.
2. Fire control programs, both from natural or human origin, must be enforced.
3. Biological particle pollution, both from animal and human origin, must be diminished.
4. Erosion control programs must be implemented, particularly in sensitive areas, such as wastelands, abandoned *chinampas* or otherwise vegetation-deprived zones.

9.5.1.3 Land

Due to the prevailing use of *chinampa* farming system in conservation soil within XT-P, water availability is critical for obtaining quality crops. Subsidence or collapse caused by water shortage will directly affect this productive system, thus reducing crop production that, although in small amounts, still reaches local markets in the southern part of MCMA.

Not all the land in the XT-P has agricultural value. Recreational activities directly related to water bodies, as well as the construction of housing complexes near them could also affect soil composition and stability. Therefore, continuous surveillance activities designed to prevent natural vegetation loss, heat build-up in the area or interference with water capture processes need to be considered to promote soil value preservation.

Management programmes for the protection of endangered or endemic species can also maintain the environmental equilibrium of the area. Because human population has been living here for quite a time, people have learned to appreciate the value of their natural resources; however, educational programmes designed to reinforce individual and collective responsibility in resource management can help to prevent biodiversity losses. From a different point of view, the XT-P could operate as a production centre for native species – trees, shrubs and even small aquatic animals with high market value – that could promote educational or ecotourism projects, representing new economic resources for the native population. This kind of endeavours, in turn, could strengthen land care.

LAND Action Guides to Drive XT-P Toward Resilient-Sustainable Scenario

1. Water recharging activities, designed to improve soil permeability and prevent its compression must be promoted, thus ensuring agricultural use.
2. Monitoring and control systems must be designed in order to continuously maintain soil quality for all kind of uses allowed within the intervention area.
3. Habitat conservation for endangered or endemic species must be promoted, through land protection programmes overseen by local inhabitants, according to up-to-date regulations.
4. Native species production projects must be supported.

9.5.2 Built Setting

9.5.2.1 Transportation

All vehicles used within XT-P must keep their emissions within the parameters set by the law, and their maintenance should be made in such a way as to assure that there is no environmental damage in the area. Parking facilities and repair shops must not be built in or near conservation areas, and in those instances when this cannot be accomplished, special precautions must be taken in order to minimize their risk of generating environmental impact. Gasoline stations must be avoided, unless they are located in avenues or freeways.

No main driveways must be built for motor vehicles within conservation soil areas, and those streets already in place must be redirected towards pedestrian use or non-motor vehicle transportation, such as bicycles. The use of motorized vehicles should be limited, when possible, to no more than 5 km-long routes, and large load trucks must be kept outside conservation areas. Garbage-collecting operations should be redesigned in order to avoid high-tonnage vehicles to enter secondary streets near the *chinampa* zone. The use of small cargo vans for garbage collection, and food and beverage transportation along secondary streets, operating in closely controlled schedules, should be stressed.

Water transportation between *chinampas* should be made using traditional paddle-powered boats, except for surveillance and emergency crafts. Whenever the use of the latter is necessary, precautions should be taken in order to keep pollution generated by them at a minimum.

Although new massive transportation systems are currently operating inside (Line 5 of the Rapid Transit Bus System, *Metrobus*) or near the XT-P (Line 12 of the Subway Metropolitan System), new bus or *microbus* routes reaching sensitive zones within this area should be avoided. Design of new transportation routes following orthogonal designs can help achieve this goal, mainly in heavily populated areas. Security issues in passenger transportation cannot be neglected. Well-lit and securely located bus stops, as well as cautiously operated and well maintained buses, must be part of an efficiently designed transportation system.

TRANSPORT Action Guides to Drive XT-P Toward Resilient-Sustainable Scenario

1. Transportation vehicles should generate the least possible environmental impacts.
2. Non-motorized vehicles (e.g. bicycles) and pedestrian roads should be promoted, especially within conservation soil areas.
3. Transportation routes must be designed, based upon primary axes, where the traffic for motorized vehicles can be speeded up, as well as upon secondary and tertiary streets, mainly designed for non-motorized vehicles.
4. Delivery routes for food and other essential goods must be reorganised in order to minimise their environmental impacts.
5. Non-conventional transportation systems must be promoted (e.g. electricity based engines, hybrid vehicles).
6. Safe and comfortable vehicles for people transportation must be made available.

9.5.2.2 Urban services

Due to the fragile nature of the XT-P, the presence of illegal settlements is a threat to resilience and sustainability. Probably the main rule to be followed in this matter should be 'No more settlements will be allowed', but there is no clear consensus regarding what to do with settlements already in place, them being legal or illegal. If we choose to keep them in place, urban services must be given to them. Old urban facilities can be a problem, particularly if they are ill operated and subject to decay by environmental factors or corruption from their operators. If possible, the replacement of aged electrical transmission lines and fractured drainage systems, as well as the closing of illegal garbage dumps should be the main focus of governmental actions. When available, new technologies of low environmental risk and impact should be taken into account.

Particularly important for the transformation of urban services are the garbage collection and disposal facilities. As previously stated, the former is probably more a transportation issue, but the latter raises the possibility of generating, directly or indirectly, health hazards for the local population. Although solid waste disposal is completely prohibited within conservation soil areas, continuous surveillance and law enforcement operations are still needed to restrain people from dumping all kinds of waste in open areas or even into water bodies. It must be stated that even crop raising and farming activities can become a problem (e.g. pesticide derived pollution, animal faeces).

People feel they may have good life quality if they hold a sense of security, a sensation usually reinforced by actions imposed by the authority. Confidence in police action must be increased, through the effectiveness of their actions, if local credibility on the governmental actions is expected. An active participation of the population through secure and anonymous reports should be promoted.

Urban planning should be an essential tool for developing actions that will allow the development of the SIMZ as a whole, and the XT-P in particular. Not only Mexico City's Government, but also local governments (i.e. *Tláhuac* and *Xochimilco* counties mayors) must define the limits of urban growth in specific areas, and should operate programmes that prevent or eliminate irregular settlements, or legalize others that can be 'rescued' through extensive urban reengineering. Social and political issues are to be taken into account when orchestrating this venture, and all stakeholders should participate seeking general agreements.

URBAN SERVICES Action Guides to Drive XT-P Toward Resilient-Sustainable Scenario

1. Maintenance and upgrading programmes for urban equipment must be created in order to maintain its efficiency.
2. Urban equipment must be designed to harmonize with proper natural resources management.
3. Non-dangerous solid waste collecting systems must be designed in order to increase their efficiency, and toxic waste generation must be prevented within the intervention area.
4. No garbage disposal final sites must be authorized within conservation soil areas.
5. Authority-enforced surveillance programmes should be reinforced while civil complaint and alert programs should be favoured, if the people's sense of security is expected to increase.
6. Urban development must be promoted only where urban equipment previously exists or where it can be harmonized with the objectives to preserve environmental quality.
7. Urban planning actions should take into account environmental, economic, social, political and cultural issues, while promoting participation of all stakeholders.

9.5.2.3 Energy

Use of conventional high-impact energy sources should be gradually reduced while seeking sustainability and resilience in any socioenvironmental system. The XT-P should be seeking elimination of electric transmission lines as a primary goal, although the high-voltage lines located near the *Periférico* probably must remain in place.

No conventional energy generation facility can be built or operated inside this area. Therefore, alternative energy generating projects should be supported, especially in areas with low population density, in order to allow solar-powered lighting in large non-wooded areas, wind-powered structures in flat lands, and biological-derived energy stations, which can be feasible due to the large amount of animal stool and stubble that is generated from agricultural activity.

On the other hand, productive activities within the intervention area, especially in conservation soil, must tend to reduce their energy consumption, whether it comes from electricity or from other conventional power sources. The employment of energy saving devices must be promoted, and storage of energy through batteries can be used whether the energy comes from electricity or from other alternative sources. This strategy must also be stressed for urban lighting systems.

Health concerns must be taken into account when building alternative energy-generating facilities. Not only environmental assessment impact studies, but also social impact evaluations must be made before any particular project can be approved.

ENERGY Action Guides to Drive XT-P Toward Resilient-Sustainable Scenario

1. Pilot programmes must be set up for solar, wind or biological driven energy generation.
2. Those products whose elaboration or uses are low energy-dependent must be governmentally promoted and stimulating their use by the population should be supported.
3. Energy saving systems must be implemented for urban lighting equipment in industrial and housing facilities.
4. No traditional energy-generating facilities must be constructed or continue operating within conservation soil areas.
5. Impact assessment evaluations must be completed, prior to allowing the construction of alternative energy generating facilities.

9.5.3 Economic Setting

XT-P can be conceptualised as a food-producing region (especially for vegetables and fruits), although its share in the market of the whole SIMZ is not large. However, its market is focused towards local consumption, and because of this, it is important that it can remain as such for sustainability and resilience purposes. Therefore, programmes must be created for boosting the consumption of local products grown in *chinampas*, while increasing their profitability through the operation of distribution networks by local population. Care must be taken that this efforts do not encourage the appearance of new farming areas that impact upon the natural environment, e.g., due to an increase in the use of inorganic fertilizers or to inappropriate employment of pesticides.

Economic interests are constantly pressuring the agricultural systems of the intervention area. Conflict of interests between those *chinampa* owners who wish to sell their land to construction companies, determined to build not only small houses for low income people, but also luxury complexes for wealthy families, and those who wish that the area can still keep its agricultural use, are constantly appearing and will surely increase through time. All stakeholders must be taken into account to reach a decision regarding this matter, and Mexico City's and local governments have an unquestionable responsibility in coordinating efforts to finally reach a consensus, also considering criteria of resilience and sustainability.

Jobs have remained a limitation for local population. Job creation program must focus on permanent, rather than temporal positions, if economic security for the people is seeked. When permanent employment cannot be obtained, complementary actions must be taken in order to temporarily alleviate people's needs. Examples of these can be the operation of communitarian support networks between producers; the creation of cooperative structures for food distribution or resource management; the capital injection by federal or local governments aimed towards resource management enterprises; and grants or funds to support producers. It must be clearly stated that, whatever strategy is set up, consensus between stakeholders as well as non-political guidance of the process, is critical for its success.

Marketing of goods produced within the XT-P has always involved middlemen and monopolisers, and nowadays the presence of big marketing chains has increased the problem. Therefore, it would be convenient to support networks that promote fair trade, initially through publicly funded initiatives or resources coming from international cooperation programmes, and later by self-supported entities that will equally divide profits between their members.

Recently, several tourism projects have been developed in conservation areas inside the XT-P. These low-impact facilities have been successful because they stress the use of culturally oriented programmes and activities, such as *temazcal* baths, handicraft making, strolls for plant and animal sightseeing or pond fishing. These activities allow local population to interact with tourists, earn their money without intermediaries, and to decide by themselves what they want to do with their venture. However, limits should be imposed on these facilities, so they do not turn into cabins or hotels with higher impacts on the environment, either by choosing to increase their host population or by turning to 'modern' activities such as motorboat touring.

ECONOMIC Action Guides to Drive XT-P Toward Resilient-Sustainable Scenario

1. Employment-generating programmes should be created. While taking into account specific features of conservation areas, they should promote wealth and allow its equitable distribution.
2. Programmes that promote economic assurance for local population must be enforced.
3. Support programmes for low-income population should be created.
4. Fair-trade systems, particularly those related to agricultural activities, should be established.
5. Projects dealing with low-impact touristic or eco-touristic activities should be governmentally supported.

9.5.4 Social Setting

Any changes that may lead towards sustainability and increase in resilience capacities must give rise to changes in values, attitudes and deep beliefs in the people involved. Consequently, educational actions (formal, non-formal or informal) should be supported, if changes in the way people use their natural resources are really hoped for. In order to be accepted, this kind of programmes should take into account the specific needs of local populations. Easily comprehended pedagogical proposals, such as workshops, shows, and ludic activities may gain the attention, particularly of children and heads of family, who can later turn them into promoters of change.

On the other hand, health status is an essential requirement for people to enjoy living. Health-promoting programmes (e.g. information and vaccination campaigns, health community clinics, epidemiological risk diminishing projects), emphasizing preventive medical actions for those considered at risk, should be stressed. While the only third-level hospital near the zone is located at the *Tláhuac* county, second-level hospitals located on easy access locations, and not only small health clinics, are needed within the intervention area. Although health care programmes are considered as high priority, proper biological hazardous materials management cannot be overlooked while operating these new health facilities.

People's interests, not necessarily related to their immediate survival, are also important in determining the perception of quality in their life. Fulfilling their intellectual, spiritual and aesthetic needs, within a tolerant environment that allows diversity of interests and respect for cultural values, is essential for people in order to consider themselves as whole human beings.

Therefore, cultural or recreational activities that could use the relaxing and peaceful environments found within the XT-P should be supported as a means of reinforcing the identity of the local population. This will allow them to express the pride of being part of the same community, encouraging them to search for common views and interests, whether or not all of them belong to the same socioeconomic stratum. As a result, shared purposes and commitments can arise, and new agreements can lead to a better coexistence between individuals.

It must be the responsibility of local authorities to invest enough resources to promote this, by supporting traditional cultural demonstrations that stem from the population's history itself, and promoting contests, theatre plays or other kinds of public performances.

SOCIAL Action Guides to Drive XT-P Toward Resilient-Sustainable Scenario

1. Education for resilience and sustainability should be promoted in order to boost the health of socioecosystems within conservation areas.
2. Health-supporting actions must be designed. These may include information and vaccination campaigns, community-health clinics and second-level hospitals.
3. Recreational activities preferred by local population should be supported, while promoting their aesthetic expressions and outdoor activities.
4. Healthy coexistence, harmony and tolerance between people, notwithstanding gender, religion or the socioeconomic stratum they belong to, should be encouraged.

9.5.5 Governance Setting

Local residents are, at least partially, descendants of people that arrived several centuries ago to the southeastern region of the Valley of Mexico, where the original population has been gradually absorbed by urban culture and its ways of living, although they still retain important elements of their original traditions.

Accordingly, actions that tend to reinforce their sense of belonging can allow them to keep their traditions, something especially important for children and young people, who are at risk of losing their love for their land, and their concern for the sustainable management of their natural resources. Examples of these programmes are the operation of teaching projects designed to understand and spread long-lasting local traditions (e.g. *El Niñopa*), craftwork making activities, traditional foods fairs, and others that seek to recover the indigenous cultures and values associated to family life within the community.

Self-esteem within a social group requires knowledge of the individual's potentialities and limitations, always within the frame of his values and deep beliefs. It will be needed that each individual within his original community, or as part of the population who has recently arrived, finds a way to reinforce his self-assurance through the recognition of his capacities, and the confidence that his opinions and thoughts can be respected and not reproached. Society as a whole, although resolutely supported by the authority, should be responsible for building the conditions that will allow this to happen. Different kinds of organizations, such as associations, clubs and action groups, can fulfil this purpose.

If the value of each individual is recognized, he/she will not fear to express his/her own opinions and share his/her ideas with the rest of the community. Anyone capable of doing so and of defending his/her ideas while respecting those of others, will also be capable of reaching consensus that will enable the group to which he/she belongs to successfully navigate through everyday life.

Such behaviours will allow people to 'empower themselves', to decide and resolve their problems, especially those related to their survival (e.g. poverty fight), their environment (e.g. sustainable use of their natural resources), their life quality (e.g. health surveillance), their connection with others (e.g. their access to information) or their self-esteem (e.g. gender equality and human rights). This empowerment should create leaders inside the organizations, who sometimes can be part of the oldest families and who can value everyone's skills and competences. Specific ways to work must be devised by each group of stakeholders, within the framework of the law and the respect to the rights of others, with their-own specific problems, without any outsider intervention.

Natural leaderships will be necessary if communication lines are to be established to allow the participatory design of viable goals for community. Even more, these lines of coordination within the community will be of great value when the government seeks their support for those objectives that have been centrally planned. Decision-making strategies for all problems that are considered important for a community must include all those interested, and meetings coordinated by acknowledged leaders should have an important part in the process.

Governance requires the existence of mechanisms that promote the settlement of conflicts within groups and organizations; the observation of equity schemes that give people the freedom to express everyone's opinions; the existence of enough tolerance to respect those who are considered different; and the possibility of constructing consensus, though not necessarily as unanimous decisions. It must be stated that not all societies reach consensus through the 'modern or globalized' visions that are now prevailing; surely those groups with traditional cultural patterns, do not.

Conflict settlement requires that people be able to make informed decisions about the problems that matter to them. Being informed is a necessary step to define the key elements that can allow a group of people to make specific reasoning and agreements that, in turn, will support decisions and actions agreed to solve the original problem.

Although information rights are essentially an individual concern, the effect that well-informed people can have on a community should not be rejected. When an individual ventures his/her opinions during a meeting, he/she can enrich the collective knowledge concerning the matter being discussed, by putting new arguments on the table.

GOVERNANCE Action Guides to Drive XT-P Toward Resilient-Sustainable Scenario

1. Self-esteem must be promoted among all members of society as a way to promote tolerance, with particular emphasis on the cultural and traditional values of local people.
2. Self-knowledge of the role of each individual within his/her community must be promoted, while encouraging his/her creativity and self-discipline.
3. Society's empowerment should be promoted, through the use of programmes that seek environmental preservation, health support, human rights observance, gender equality, poverty combat and general access for information and communication technologies.
4. Programmes that promote creativity, teamwork and leadership as catalysers for change must be supported.
5. All social leaderships, as well as cultural values, must be respected while dealing with local cultures.
6. Conflict-settlement mechanisms should be promoted.
7. Information rights must be protected, regarding all activities that can influence life-quality perception by all people.

9.6 Concluding Remarks

For nearly seven centuries the Valley of Mexico has been subject to anthropogenic changes that have gradually transformed its lakeside landscape to make it one of the 10 most populated and large metropolitan areas on the planet. A transformation of this magnitude could not happen without altering the natural and human balances that support the resilience and sustainability of the area. To complete this difficult scenario, it is necessary to remember that many years ago, MCMA lost its water, food, energy and waste management self-sufficiency. Multiple goods and services of the MCMA have to be 'imported', thus reducing the resilience and sustainability of its supply regions, with the consequent transfer or externalization of environmental and social costs.

There are multiple interests, resistance and inertia that the current development model in Mexico, focused on capital growth, imposes to the task of recovering sustainability and improving resilience for the MCMA. The fundamental questions to be answered are: What notions and approaches of resilience and sustainability will guide this transformation? Which changes are we willing to support and which others we are not? Are we determined to face the cost and restrictions that can result from this transformation?

The XT-P faces strong internal and external pressures due to its population dynamic; it is a transition zone between rural and urban media that has altered both their population's original culture and lifestyle, with traditional production systems losing competitiveness against current trends in production and marketing. Thus, it would seem that if the culture and tradition of the zone are to be preserved, it would be necessary to reorient this production system (e.g., adding value through organic certification or floriculture) to make it viable, profitable, resilient and sustainable.

The prevailing economic dynamics, as well as the struggle for political control of the region, has led to a weakening of social networks that used to give cohesion and strength to the community. There is no recognition (or compensation) on environmental services that the XT-P provides to the MCMA; water supply management policies based on demand have ignored the natural water cycle.

The deterioration of the social and institutional structure within the XT-P is what ultimately is undermining the environmental integrity and resilience of the area. In this way the socioecological deterioration is endangering the resilience and sustainability of one of the last bastions of wetlands within the Valley of Mexico, recognized by UNESCO as cultural heritage of mankind.

If current trends continue within the XT-P we could reach a stage in which:

- Urban projects would tend to spread at the expense of conservation areas that provide much-needed environmental services to the Valley of Mexico.
- Wetlands would disappear and with them the associated biodiversity.
- Landscape characterized by water mirrors, *chinampas*, *ahuejotes*, and background volcanoes would become a very grey and monotonous landscape of urban developments.
- The area would become a new heat island in the south of the Valley of Mexico, thereby causing changes in wind, rainfall and temperature patterns.
- As a result of the above, the air pollution problems in the Valley of Mexico would increase.
- Socioeconomic and cultural matrices of the area would yield to the pressures of urbanization, thereby losing its historical heritage and irreplaceable identity (pride of Mexico and the world).
- The possibility of exploiting the potential tourism, as alternative to local development would be depleted.

For decades, stakeholders have implemented several measures to address environmental and social issues that are emerging, but there is much more that can be done to recover resilience and sustainability within the MCMA. The target image proposed in this paper incorporates a vision in which the socioecological systems within the XT-P establish a new state of dynamic equilibrium in which:

- The *chinampa* area at least maintains its current extension and recovers its strength as a development alternative for the inhabitants of the area.
- XT-P becomes self-sufficient in water resource management, thereby strengthening the resilience of the ecosystem. In this way the wetland area within the polygon will decrease its vulnerability.
- Production systems within the XT-P recover their self-sufficiency in their treatment of solid and liquid wastes.

- The sense of identity and pride of the people (especially young people) is strengthened, recovering, enriching and resignifying the manifestations of their local culture.
- Productive activities are diversified (e.g. floriculture, organic farming, tourism in contact with nature, scientific tourism).
- The quality of orderly and responsible touristic services constantly increases XT-P's international recognition, and its cultural heritage is consolidated.
- At the XT Polygon:
  - While maintaining a position of openness to interact with other stakeholders, local communities consolidate self-management processes through which they identify, prioritize, seek alternatives and solve their own development problems.
  - The education system (formal and informal), as well as the citizens, assume their commitment in the process of forging a culture in which the characteristic values that promote sustainability emerge (e.g. responsibility, prudence, pluralism, respect, inclusion, solidarity, collaboration, equity, justice, forethought, proactivity, austerity).
  - Communication processes continually improve by promoting access to information, assertiveness, empowerment, empathy and agreed resolution of conflicts.
  - Civil society initiatives arise to create 'lighthouses' that continuously monitor, evaluate, inform and encourage progress towards resilience and sustainability.
  - A high level of governance is reached.

From a methodological point of view, this study shows the relevance that for any intervention project has the use of a 'powerful gradient' between the description of the current problematic situation (diagnosis) and the objective image or idealized (prospective) situation that is designed in order to build an action guide for the system subject to study-intervention programmes.

Finally, it should be noted that the set of descriptors and action guides proposed by the Interdisciplinary Sustainability Team are a conceptualization step in an effort to construct viable indicators (qualitative and quantitative) that seek to evaluate the status and progress of resilience capacities and sustainability in the XT-P.

References

Checkland, P. (1997). *Pensamiento de Sistemas, Práctica de Sistemas*. Ed. Noriega, 1ª reimpresión de la versión autorizada en español de la obra publicada en inglés en 1981.

Clayton, A., & Radcliffe, N. (1996). *Sustainability: A systems approach*. Earthscan.

Ezcurra, E. (2000). *El Ecosistema Urbano*. In: G. Garza (coord). La Ciudad de México en el final del Segundo Milenio. El Colegio de México.

Garza, G. (2000). *Introducción*. In G. Garza (coord). La Ciudad de México en el final del Segundo Milenio. El Colegio de México.

INEGI. (2015). *Cuaderno Estadístico Delegacional de Xochimilco y Tláhuac*. GDF.

International Centre For Sustainable Cities (ICSC). (2008). *The sustainable cities: PLUS planning cycle of Calgary, Canada*. Retrieved from http://www.icsc.ca/content/dmdocuments/plus_planning_cycle.pdf

López, L. A. (2004). *La chinampa como patrimonio cultural: identificación de problemas y alternativas de manejo. Informe de resultados del taller con chinamperos del pueblo de San Luis Tlaxialtemalco*. UNESCO-Xochimilco.

Mazari, M., Noyola, A., (2000). *Contaminación del Agua*. In G. Garza (coord). La Ciudad de México en el final del Segundo Milenio. El Colegio de México.

Pisanty, I. (2000). *Ecosistemas y Áreas Verdes* In G. Garza (coord). La Ciudad de México en el final del Segundo Milenio. El Colegio de México.

Secretaría de Desarrollo Urbano y Vivienda (SEDUVI) and Universidad Autónoma Metropolitana (UAM). (2008). *Polígono Xochimilco-Tláhuac en el Marco del Programa de Desarrollo de la Zona Integral Metropolitana Sustentable. Anexo Mesa 2: Propuesta de Descriptores para el Análisis de Sustentabilidad en Diferentes Sistemas de Estudio*. SEDUVI.

United Nations Educational, Scientific and Cultural Organization (UNESCO)-Gobierno del Distrito Federal (GDF)-Delegación Xochimilco. (2006). *Xochimilco. Un proceso de gestión participativa*. UNESCO-Xochimilco.

United Nations Environment Programme (UNEP) and Centro de Investigación en Geografía y Geomática "Ingeniero Jorge L Tamayo" A.C. (Centro GEO). (2004). *GEO Ciudad de México, México*.

Universidad Autónoma Metropolitana (UAM). (2006). *El Plan Institucional hacia la Sustentabilidad (Pihasu)*. UAM.

Chapter 10
Eco-neighborhoods as a Way to Strengthen Urban Resilience

Alberto Cedeño-Valdiviezo

Abstract Eco-neighborhoods – ecological settlements that began to emerge in the 1990s with the global impetus of sustainable development – are presented as a contribution to urban resilience, so that their knowledge and the possibility of Its reproduction. This article explores what makes a city habitable and sustainable, defines eco-neighborhoods, presents their objectives, and describes some examples in several nations.

Keywords Eco-neighborhood · Sustainable neighborhood · Sustainable development · Habitable city · Sustainable city

10.1 Introduction

10.1.1 The Sustainable Livable City

The quality of life of Mexico City residents has gradually but continually declined. Currently they live with high levels of air contamination and face enormous risks of an earthquake such as that which occurred 1985 as well as an airport and sand mines in very densely populated areas, in addition to increasing levels of insecurity. Given the incapacity of past and present administrations to resolve these problems and the fact that large numbers of people have migrated to this city over the past five decades, there is a critical need to study urban resilience in Mexico City and evaluate the city's capacity for overcoming these risks.

Many Mexico City residents question what may be done to improve the city's quality of life. As the neighborhood has been a very important social unit for the same, one interesting answer is to promote development of eco-neighborhoods – urban ecological settlements – as have arisen in several countries since the 1990s.

A. Cedeño-Valdiviezo (✉)
Autonomous Metropolitan University, Mexico City, Mexico

O. F. González Castillo et al. (eds.), *Urban Resilience: Methodologies, Tools and Evaluation*, Resilient Cities, https://doi.org/10.1007/978-3-031-07586-5_10

Eco-neighborhoods could provide an example for the local government to transform the paternalistic approach with which it administers the city, as well as the mentality of the inhabitants the greater metropolitan area.

As early as 1961 in her *Death and Life of Great American Cities*, Jane Jacobs wrote about the mistakes of modern urbanism that has turned the neighborhoods of cheap housing into worse centers of delinquency, vandalism and social despair than the slums they were to replace (many of which are existing buildings since many years) and states that in the motivations of crime – both in the outlying districts and in provincial towns and in large capitals – there is undoubtedly a substratum of deep and complicated social pressures (Jacobs, 1961). It is these social pressures on which we have to work, and we believe that only by achieving groups that work as teams can be achieved. Eco-neighborhood represent a solution for the possibility of a sustainable livable city.

For Jane Jacobs, the elements that define a proper neighborhood should be:

- There is no direct relationship between a good house and good behavior of the neighborhood residents.
- It is not up to good schools to rescuing a neighborhood.
- Neither is the socioeconomic condition of its inhabitants.
- It is very important a good self-management, for which we must rid ourselves of any ideal neighborhood as introverted and self-sufficient unit, since we can't forget that they are part of a city.
- Good neighborhoods are not discrete units. They are physical, social and economic continuities, on a small scale of course.
- Interesting and lively streets should be promoted.
- Connecting the streets with the effective organism of the district, and making the fabric of these streets as close to a continuous network.
- Importance of developing the district as a district operating on a larger scale than agglomerating smaller neighborhoods, and in which there must be an efficient administration.
- Emphasize the functional identity of areas large enough to operate as districts.
- Make parks, squares and public buildings part of the street fabric. Avoid at all costs the creation of different islets in their activities and functions.
- The size of the functionally viable district will depend on the size of the city as a whole (Jacobs, 1961).

In 2010, the local government of the Mexico City neighborhood Coyoacan had hoped to develop something similar to an eco-neighborhood (La Jornada, 2010), although this did not come to fruition. That year, they organized the First Workshop on Development of Eco-neighborhoods, which included a visit to the Eco-village of Huehuecoyotl (Caravana Arcoiris por la Paz, 2012). Eco-villages were the rural precedent of urban eco-neighborhoods. Huehuecoyotl was started in 1982 in the mountains of the Sierra de Tepoztlan, close to the village of Santo Domingo Ocotitlán (Hieronimi, n.d.). In this eco-village, people of several nationalities research and model a lifestyle oriented toward individual and community well-being and sustainability based on ecology, the arts, true democracy, and holistic health practices

(Aldea Ecológica Huehuecoyotl, n.d.). Residents have subsisted principally by selling handcrafts and holding permaculture courses (Hieronimi, n.d.). While Mexican government officials do not foment such settlements, ordinary citizens have organized to do so, but examples are few and far between. Before looking into how to create such communities, we should ask what makes a city habitable.

Esther Higueras (n.d.) states that livable cities should adapt their layout to make all facilities and services necessary for its inhabitants accessible by walking; provide sufficient public green spaces and incentives for ecologically sustainable commercial activities; and be safe, friendly, and healthy – contamination-free without excess noise. This author considers that livable, sustainable cities require:

- Environmental planning
- Bioclimatic urbanism (explained below)
- Environmental laws and guidelines
- Ecological and bioclimatic rehabilitation of existing buildings and infrastructure
- Long range city planning according to the UN voluntary action plan "Agenda 21".

Higueras adds that environmental indicators should be used to continually evaluate whether a city is livable and meets sustainability objectives; such indicators measure: water quality, air quality, noise level, traffic safety, housing quality, and accessibility of green spaces.

As Higueras (n.d.) mentions, territorial environmental planning is essential for mid- and long-term regulation in order to achieve sustainability and prevent negative impacts. It is important to identify the possible nature and scope of potential impacts of possible management strategies in order to select those which are least harmful to the environment and establish preventative and corrective measures in accordance with the local context. For example, land use changes which could negatively impact the environment should be prevented. Land use by residents and tourist should be determined by the territory's carrying capacity, which defines limits to land use and establishes equilibrium with the natural environment. Such planning must include an environmental impact study.

Another key element to environmental planning is bioclimatic urbanism, which establishes rules and standards regarding urban growth and architecture considering the territory, including its climate, and landscape. Bioclimatic urbanism takes into account the sun and wind, land use, construction materials and techniques, population densities, and the relationship between municipal facilities and housing, and foments clean renewable energies at the building and urban levels (Higueras, n.d.).

Laws and regulations which establish the above-mentioned conditions for urban areas may lead to design of buildings and urban spaces which are more energy efficient and make more efficient use of materials, thus avoiding unsustainable construction and urban layout (Higueras, n.d.).

With respect to ecological and bioclimatic rehabilitation of an existing city, rehabilitating a building leads to energy savings of 60% as compared to demolishing and rebuilding, and avoids numerous environmental impacts. Three aspects of rehabilitation should be considered: sustainable rehabilitation consists of concrete actions which foment social, economic, and environmental sustainability; ecological rehabilitation seeks to close urban ecological cycles of water, organic matter, energy, and waste; and bioclimatic rehabilitation aims to reduce energy consumption through passive heating and cooling techniques in buildings and open spaces (Higueras, n.d.).

Finally, long range city planning based on sustainability criteria according to the UN voluntary action plan "Agenda 21" will help guide a municipality's philosophy, strategy and operational programs toward improving the life of its citizens.

10.2 Eco-neighborhoods

This section will explain what eco-neighborhoods are considered to be and how the idea originated. In 1960, Kevin Lynch defined neighborhoods in general as relatively large urban zones whose defining characteristic is that they are recognizable and may be used as a reference point when one is outside of them (Lynch, 1984).

Jacobs refers to the districts as harmful to urbanism, since it impels to model to force the life of a city like imitations of the residential or provincial life. He adds that sentimentality plays with mellow intentions rather than common sense. It also refers to the neighborhood well succeeded as a place that keeps its problems at a distance such that it is not allowed to destroy by them (Jacobs, 1961)

While the neighborhood as a social entity in the United States (Lynch's and Jacobs's country of origin) often has little social significance, in Latin America the neighborhood is practically the "homeland" of those who live in it. Therefore, in Mexico, in order to increase social cohesion which, could help avoid many social problems, it is important to strengthen these urban spaces.

Verdaguer (2000) states that traditional urbanism focuses on the neighborhood scale, its facilities, natural resources, and the general structure of the city. This author adds that in a well-integrated eco-neighborhood, residents feel part of their environment, and the neighborhood has a reciprocal relationship with surrounding neighborhoods, outlying areas, and farther beyond, as well as access to public services and facilities. Other distinctive traits include respect for pre-existing landmarks which local residents consider to be signs of their local identify, and respect for and integration of natural and human elements of the urban landscape (Verdaguer, 2000).

These considerations by Verdaguer are particularly significant given the times in which they were proposed – when eco- or sustainable urbanism and eco-neighborhoods were just beginning to be considered. Since then, eco-neighborhoods have become paradigms of sustainability.

Some of the earliest authors to mention eco-neighborhoods were Rudin and Falk (1999), in *Building the 21st century home*, and later Cuello (2012), in *The Sustainable Urban Neighbourhood*. However, practical examples existed before the concept was coined. One of the first known eco-neighborhoods was the Puchenau Garden City in Linz, Austria, whose construction began in 1962 (Ruano, 1999).

10.2.1 Defining an Eco-neighborhood

As the concept of eco-neighborhood is recent, it lacks a widely agreed-upon definition, and its scope and limits are still not clear. Based on traditional definitions of neighborhood, Cuello (2012) describes eco-neighborhoods as urban neighborhoods with a high level of environmentally, socially, and economically sustainable development, which are rooted in the discipline of urban development as well as in the concepts of re-development, urban renovation, and holistic urban improvement. This author adds that an eco-neighborhood is a community of people that with a common long-range vision organizes with the aim of improving their quality of life and achieving human well-being in harmony with the environment. An eco-neighborhood follows the principals of eco-villages – those human settlements that seek sustainability of all aspects of life, including the physical environment as well as economic, social, and cultural aspects, in harmony and with respect for the natural environment, promoting healthy long-lasting systems of development. Eco-villages are generally considered to be those rural communities with such a focus, while eco-neighborhoods generally refer to such urban communities (Organi-K, n.d.), although some authors such as Cuello (2012) consider that eco-villages may also be urban.

Higueras (n.d.) defines an eco-neighborhood as a segment of a city designed to meet requirements of energy efficiency, for example by implementing active and passive heating and cooling systems, in order to establish an "adequate relationship with its environment".

Given that as of yet the term eco-neighborhood has no official universal definition, it may be conceptually confused with the term eco-city; however, the eco-neighborhood is part of an existing city. Eco-neighborhoods are not limited to new developments, but may also include recovery and rehabilitation of existing urban buildings and open spaces. Eco-neighborhoods also prioritize social and economic aspects such as diversity and citizen participation, collective management of community goods and services, local development, and lifestyle including diet, consumption of other goods, creativity, and use of time. From a perspective of complexity, all these aspects may mutually influence each other in a synergetic manner (Morán, 2008).

10.2.2 Elements of an Eco-neighborhood

According to Rudin and Falk (1999), a sustainable urban neighborhood must follow four basic principles:

- **Environmental sustainability:** This includes land recovery, and addresses population density, transportation, energy efficiency, waste, the water cycle, and green spaces.

- **Sustainable urban structure:** This takes into account streets, blocks, landmarks, and other public spaces, as well as communications, and other networks. Urban structure should be dense, compact, and varied, and promote a mixture of uses. Municipal authorities should promote exchanges with other neighborhoods to share their model and learn from others.
- **Social sustainability:** The neighborhood should be characterized by complexity, with a mixture of uses and housing types. Residents should have common interests and objectives, a sense of appropriation and responsibility, and participate in design and management of community facilities and systems.
- **Natural growth:** The city should be allowed to evolve in an organic manner; participation of different urban designers and architects promotes diversity (Cuello, 2012).

Verdaguer (2000) mentions the following as being important to sustainable urban neighborhoods:

- Compactness
- Mixture and flexibility of uses
- Integration of nature, cultural landmarks, and other aspects of the landscape in urban spaces
- Public space being important to citizens' daily life
- Bioclimatic construction
- Recovery of buildings and infrastructure
- Sustainable transportation
- Renewable energy system

Rueda Palenzuela (2013) considers the definition of eco-neighborhood to include design, construction, and/or remodeling of neighborhoods according to four fundamental principles, which differ from those proposed by Rudin and Falk:

- **Compactness:** This facilitates contact, communication, and exchanges among residents.
- **Complexity:** Urban spaces should include a mixture of uses and functions, which promotes residents' creativity.
- **Efficiency:** The neighborhood should make wise use of resources and cause minimal disturbance to ecosystems.
- **Social stability:** This promotes social diversity and cohesion, necessary for democracy and equal opportunities.

Morán (2008) presents a more complete position based on three principal requirements similar to those named by Rudin and Falk:

- **Environmental sustainability:** This involves closing cycles of energy and matter with respect to territorial use, including transportation as well as urban metabolism.
- **Sustainable urban model:** The physical design, economic activities, and the growth process should be sustainable.

- **Social sustainability**: Eco-neighborhoods should promote harmonious social networks and relationships, as well as responsible public administration.

Morán considers that **environmental sustainability** requires taking into account the urban metabolism (energy, transportation, water, and other materials), as well as the neighborhood's insertion into the built and natural environments, prioritizing reuse of already developed land and built patrimony over new development, and following the hierarchy of "reduce, re-use, recycle". With respect to the **urban model**, the neighborhood should be autonomous and at the same time connected to outlying areas, and dense with a mixture of uses – residential, employment, and services – with short distances among these areas. Attention should be paid to endogenous economic development, public spaces, and modes of urban growth. Finally, Morán considers **social sustainability** to mean that the neighborhood provides a place for social encounters and participation in community life. It must be taken into account that the neighborhood is a complex organization, and as a complex entity it accumulates information and thereby has the capacity for adaptation and transformation.

10.2.3 Objectives of Eco-neighborhoods

According to Hernández et al. (2009), the environmental objectives of an eco-neighborhood are:

- Closing the water cycle
- Savings in energy, water, and materials
- Improving natural areas and biodiversity
- Working with the climate
- Creating an attractive living and working environment
- Establishing short distances between services and living and working spaces, and ease of movement among them
- Providing basic services and infrastructure to facilitate daily life for current and future inhabitants

These authors consider that eco-neighborhoods should seek to comply with the following principles:

- Integration among residents
- Social responsibility
- Diversity
- Respect for nature
- Mobility
- Urban metabolism
- Sustainable construction

The digital magazine Organi-K, also hold that eco-neighborhoods should carry out the following:

- Diagnostic study of environmental impact on urban or peri-urban land, and/ or on conservation land
- Diagnostic study and selection of the target population for which eco-neighborhood strategies and techniques will be implemented
- Creation of grassroots organizations in which members of an eco-neighborhood may participate in decision making as well as social and political activities
- Training of community members in facilitation and consensus techniques
- Document agreed upon by the community establishing agreements and commitments
- Acquisition of an environmental culture and sustainable habits
- Construction using ecological techniques
- Following natural architecture principles when new homes must be built
- Maintenance of existing ecological technologies in homes
- Monitoring and follow up of environmental impact (Organi-K, n.d.).

According to Higueras (n.d.), eco-neighborhoods should:

- Close matter and energy cycles
- Establish a sustainable relationship with their territory and landscape
- Reduce air, water, and soil contamination
- Improve residents' social relationships

For this, the basic conditions of eco-neighborhoods are:

- Linking aspects of the urban structure: The eco-neighborhood should be part of the structure of an existing city, providing it with alternatives oriented toward efficiency.
- Rehabilitation and re-use of built patrimony
- Compactness
- Mixture of uses: Economic, recreational, and social spaces should be designed so as to provide maximum flexibility in spatial and temporal use of the neighborhood.
- Social cohesion and participation: Citizens should be aware of the condition of their city and how they may improve it. For this, it can be helpful to follow the UN voluntary action plan "Agenda 21".
- Habitability: high quality of life standards
- Conception of public space as important to citizen life
- Inclusion of nature in the city: green spaces that foment health
- Predominance of public transportation, biking, and walking over private vehicles
- Use of clean renewable energies on a municipal scale and in homes
- Management of water demand: education and consciousness raising regarding the value of water and implementation of strategies to maximize re-use
- Reduction, re-use, and recycling of urban waste

One of the most basic principles in designing eco-neighborhoods is rehabilitating and re-using built patrimony. Before considering re-building, it is important to evaluate the possibility of rehabilitating and re-using existing constructions (Higueras, n.d.).

Higueras (n.d.) proposes the following general criteria for designing an eco-neighborhood:

- Establish a hierarchy of public spaces divided into principal and secondary spaces which are connected in a network by streets or paths
- Adequate design of streets and parks
- Safety and security in public spaces
- Consider the needs and social conditions of those who will use these spaces upon designing layout
- Use of legible signage with easily recognizable symbols on signs and monuments to promote group identity
- Preserve each site's uniqueness
- Connect public spaces to infrastructure, establishing appropriate distances for accessibility to services, and using appropriate street furniture

10.2.4 Are Current Eco-neighborhoods Really Ecological?

Paco Segura from Ecologists in Action and and Evangelina Nucete from WWF Spain (personal communication) consider that publicity regarding eco-neighborhoods could be exaggerating their success. While buildings incorporate bioclimatic architecture which increases insulation and include thermal and/or photovoltaic solar panels or other energy efficient energy mechanisms (heating with biomass, district heating, etc.), such techniques are not new, and in Brussels, for example, zero energy homes are being planned (Casa Bioclimática, 2009).

With respect to negative aspects of these "ecological neighborhoods", Segura and Nucete hold that "all of them suffer from a very strong deficiency: transportation, today one of the principal energy consumers". Many of these new eco-neighborhoods are proposed on city outskirts, far from urban centers; despite having bike lanes and pedestrian zones, their residents generally depend on private vehicles. Furthermore, eco-neighborhoods tend to involve new constructions. Nucete recommends that regional governments focus on rehabilitating existing neighborhoods – particularly degraded city centers – according to sustainability criteria (Casa Bioclimática, 2009).

10.2.5 *Ecobarrios Throughout the World*

In many nations – Germany, Australia, Austria, Denmark, United States, Finland, Greece, England, Mexico, and Sweden – eco-neighborhoods were developed before the 1992 Rio Summit. Even before the 1987 Brundtland Commission report, which popularized the concept of Sustainable Development, eco-neighborhoods were reported in Austria, United States, Greece, and Mexico; most were probably inspired by the 1972 Stockholm Earth Summit.

Two of the first known eco-neighborhoods are the Puchenau Garden City in Linz, Austria, whose construction began in 1962 – even prior to the 1972 Earth Summit, and Solar Village No. 3 in Pefki-Lykovryssi, Athens, Greece, which began in 1978 (Ruano, 1999).

Several eco-neighborhoods exist in France. In the city of Doaui, the eco-neighborhood "Le ZAC du Raquet" was started in 2006 to house 12,000 inhabitants. In Merville, "Les Jardins de Flanders" is planned to include 350 homes built according to ecological criteria. In Courcelles, "Le Domaine de la Marlière" is being developed for some 1230 homes. Strasbourg's municipal administration recently announced construction of several ecological neighborhoods (Eroski Consumer, n.d.).

There is a need to carry out a current review of the world's most important eco-neighborhoods. Below, we describe a few of the most well known eco-neighborhoods.

10.2.5.1 Viikki, Helsinki, Finland

Construction of the Viikki Eco-Community Project (Korpivaara, 2005) began in the 1990s, ending in 2010. Viikki offers a welcoming place for 1700 people to live. Community members coordinated the construction with Finland's Environmental Ministry, the national agency for technology, and an architects' association. Homes are built with bio-construction techniques in order to incorporate many types of active and passive bio-energy systems and green spaces and take advantage of rainwater (Casa Bioclimática, 2009).

Buildings were designed to take maximum advantage of sunlight and avoid projecting shade on the other buildings. Vegetation was distributed to reduce exposure to predominant winds, and buildings are lower than surrounding natural barriers. Private gardens and public green spaces between buildings provide pleasant surroundings and allow for drainage and recovery of rainwater through a community-wide capture system (Cátedra Municipios Sostenibles, 2011).

The Viikki complex was designed with an open plaza bordered to the south by adjoining two-story homes protecting it from the dominant wind, to the north by a four story building, and to the east by a small laundry room and other common spaces. Pre-fabricated materials were used maximize quality and reduce construction waste. For example, pre-fabricated concrete beams and posts come with insulation already integrated (Cátedra Municipios Sostenibles, 2011).

Climatic comfort and energy savings has been achieved thanks to the high thermal mass, super-isolation, use of grey and white on the facades to regulate heat absorption, greenhouses incorporated into the south sides, use of double low emission glass, a heating system using hoses in the ground, and 63 solar collectors that heat 60% of the hot water used in the community. Furthermore, an air circulation system heats air from the north in the summer and from the south in the winter by passing through the greenhouses (Cátedra Municipios Sostenibles, 2011).

We mentioned that one critique of eco-neighborhood is planning and management of transportation and traffic, which in this eco-neighborhood was considered in the territorial arrangement. In the Viikki community, all modes of transportation are used: public transportation, private vehicles, biking, and walking, facilitated by pedestrian networks (Cátedra Municipios Sostenibles, 2011).

10.2.5.2 Vauban in Freiburg, Germany

According to members of Ecologists in Action and WWF Spain (personal communication), the Vauban eco-neighborhood, located in the German city of Freiburg, is a paradigm for how authentic eco-neighborhoods should be established. "Children may play safely in the streets; 40% of residents do not have a car and those that do park in a communal zone in one end of the neighborhood. Furthermore, Freiburg's optimal system of public transportation facilitates residents' mobility (Observatorio de Ciudades Inclusivas, n.d.).

The city of Freiburg, with approximately 250,000 inhabitants consists of 15,300 ha (including 6533 ha of forest) and is close to the Black Forest and the Swiss border (Observatorio de Ciudades Inclusivas, n.d.). The Vauban neighborhood dates back to 1937, when barracks were built in the outskirts of Freiburg to house Hitler's Wehrmacht troops. Once the Second World War ended, the military zone was confiscated by the French army, which established a NATO base there, called the Vauban Quarter. In 1992, after German reunification, the French troops withdrew and the barracks were abandoned (Fernández et al., 2010).

Freiburg's commitment to caring for the environment originated with environmental protests of the 1970s, which were fundamentally anti-nuclear, and ecological parties have governed this city for decades. Freiburg has worked toward sustainable transportation, with 400 km of bike lanes and a streetcar system, and has installed renewable energy systems in numerous public buildings, offices, homes, and places of worship (Fernández et al., 2010).

The Freiburg City Council acquired the Vauban Quarter land from the federal government in 1993 in order to develop a residential neighborhood; with community participation, this became an eco-neighborhood (Morán, 2008).

Besides enjoying a sustainable home, the almost 5000 residents of the Vauban eco-neighborhood have access to the community facilities (such as a hospital, school, and shops), green spaces, a variety of social and cultural activities, and social and community networks, all of which promote social cohesion. The project's promotional literature points out that the community context is particularly adequate for childhood development. Other beneficiaries include farmers and producers of other ecological products that sell to community members and in their weekly open market (Observatorio de Ciudades Inclusivas, n.d.).

Current and potential residents have actively participated in environmental urban renewal, including meetings and workshops led by other people knowledgeable of topics such as alternative health, construction, and transportation, as well as community development with the aim of emphasizing local rootedness. For example, in 1997 and 1998 efforts were geared toward researching possible alternatives with respect to bioclimatic construction, sustainable mobility, and energy efficiency (Observatorio de Ciudades inclusivas, n.d.).

The initiatives known as "neighborhood without cars" and "neighborhood without parking" prioritize walking and biking. No parking areas may be built next to the homes; as stated above, parking is limited to one end of the residential area, and within the residential area use of cars is limited to medical or other emergencies, pick-up, and drop-off at 5km/h; 30km/h is allowed in the rest of the community. Services, workplaces, and other facilities within the neighborhood are accessible by bike or walking. Streetcars, buses, and trains connect Vauban with the center of Freiburg. Close to half the residents do not own a car, although a community-wide association of shared cars and vans exists (Observatorio de Ciudades Inclusivas, n.d.).

Efforts have been made to respect the natural water cycle. Stream edges have been conserved, and the neighborhood has a sewage and grey water filtering system. Thanks to the abundance of green spaces, 80% of the residential area permits rain filtration. A pilot project of vacuum toilets directs sewage and organic waste to a biogas plant to produce cooking fuel for several homes. Grey water is purified in a biological filtering system and used for watering or otherwise returned to the ground (Morán, 2008). Septuagenarian trees have been conserved, and upon building, residents are the obligated to conserve existing trees and plant others, use ecological construction materials, include green rooves and rain filtration systems in their homes, and consume less than 65 kWh/m^2 annually; use of solar energy through photovoltaic collectors and panels is recommended (Morán, 2008).

The Vauban eco-neighborhood's most significant problem is that while efforts have been made to attract a diverse population, the existing population is quite homogenous – principally middle class university students, especially since most previously existing low-income municipal rent subsidies were eliminated (Observatorio de Ciudades Inclusivas, n.d.).

10.2.5.3 BedZED in England

Fifteen kilometers from London, "BedZED" (Beddington Zed Energy Development), a 3000 m^2 residential complex with 92 homes built from 2000 to 2002, is the most complete eco-neighborhood in the United Kingdom. The project led by architect Bill Dunster is characterized by zero fossil fuel consumption, use of bio-climatic energy techniques, recycled materials, responsible water use, and – most interestingly – conversion of an old waste treatment plant into housing (Eroski Consumer, n.d.).

BedZED has been visited by thousands of people, and is the most ambitious example of novel sustainable techniques and practices in a new housing development. The fact that it is in a popular, attractive part of the metropolitan area demonstrates that sustainable living does not have to be a sacrifice or lack comfort. The children play happily and safely in its pedestrian streets. Payments for electricity and heating, as well as housing prices are lower than for other homes in the area (Biorregional, n.d.).

The design of the ecological homes facilitates heating and cooling and incorporates rainwater collection, while the design of the neighborhood as a whole foments pedestrian mobility over use of automobiles and includes pedestrian streets and convenient bus stops to minimize environmental impact (Arcia, 2012).

10.3 Conclusions

Achieving a high quality of life according to principles of sustainability and urban resilience is not easy in today's cities, especially given their limited space and large size. We have provided examples from Europe in which municipal administrations have worked with citizens´ groups to improve their cities' quality of life and the result have been the eco-neighborhood. In Mexico, we are still far from this possibility. There is a need for municipal authorities and inhabitants to learn to work together toward this objective, and it is important to learn to live in a sustainable way. This article shows that eco-neighborhoods in Europe have managed to achieve this. However, doing so requires that inhabitants have a high level of consciousness, as well as environmental education, which is lagging in Mexico.

To show what these human settlements consist of, we have defined eco-neighborhoods, their objectives, and their principle characteristics; though their defining characteristics are still not very clear, practical examples are gradually filling in the gaps, showing that eco-neighborhoods are a good model for Mexico's cities. Specialists point out that although eco-neighborhoods use eco-techniques and bioclimatism, they still consume energy and lack sustainable, efficient transportation systems. Those existing eco-neighborhoods in several countries – in particular Germany, Finland, Switzerland, and England – have resulted from inhabitants' interest in transforming their way of life. Other countries are gradually following suit, providing an example for Mexico City, which is in need of fundamentally improving its quality of life, since it is currently terrible.

References

Aldea Ecológica Huehuecoyotl. (n.d.). http://huehuecoyotl.net. Accessed 18 July 2016.

Arcia, M. E. (2012). *Un barrio de casas ecológicas en Londres*. http://icasasecologicas.com/. Accessed 29 Oct 2012.

Biorregional. (n.d.). *The BedZED Story*. www.bioregional.com/wp-content/uploads/2016/04/The-BedZED-Story.pdf. Accessed 18 July 2016.

Camacho, F. (2010, June 6). Ecobarrios, propuesta ante un panorama donde las megalópolis se colapsan. *La Jornada*. Mexico City. http://www.jornada.unam.mx/2015/06/06/cultura/a03n1cul. Accessed 16 Aug 2016.

Caravana Arcoiris por la Paz. (2012). Ecoaldea Huehuecoyotl, Mexico City. http://caravanaarcoiris.blogspot.mx/. Accessed 16 Aug 2016.

Casa Bioclimática. (2009). Barrios ecológicos. http://www.casabioclimatica.com/noticias/blog/casa-bioclimatica/barrios-ecologicos/. Accessed 20 Aug 2016.

Cátedra Municipios Sostenibles. (2011). *Barrio de Viikki, ciudad de Helsinki, Finlandia*. Universidad Politécnica de Valencia y la Red de Municipios Valencianos hacia la Sostenibilidad http://camuniso.blogspot.mx/2011/01/barrio-de-viikki-ciudad-de-helsinki.html. Accessed 19 July 2016.

Cuello, J. F. (2012). *La verdad sobre los ecobarrios*. http://urbanismosostenible.blogspot.mx/2012/08/la-verdad-sobre-los-ecobarrios.html. Accessed 13 July 2016.

Eroski Consumer. (n.d.). *Barrios ecológicos*. www.consumer.es/web/es/medio_ambiente/urbano/2009/03/18/184097.php?page=3. Accessed 18 July 2016.

Fernández, J. L., Ramos, A., & Morán, N. (2010). De cuartel militar a laboratorio de vida alternativa: el ecobarrio de Vauban en Friburgo. *Electronic journal: Papeles de relaciones ecosociales y cambio global*. E.T.S. Arquitectura, Universidad Politécnica de Madrid. https://raicesyasfalto.wordpress.com/2015/02/01/de-cuartel-militar-a-laboratorio-de-vida-alternativa-el-ecobarrio-de-vauban/. Accessed 19 July 2016.

Hernández, A. A., Velázquez, I., & Verdaguer, C. (2009). *Ecobarrios para ciudades mejores*. Universidad Politécnica de Madrid.

Hieronimi, H. (n.d.). *Huehuecoyotl – una ecoaldea en Morelos/ México*. tierramor.org. http://tierramor.org/Articulos/huehuearticulo.htm. Accessed 9 Jan 2017.

Higueras, E. (n.d.). *El reto de la ciudad habitable y sostenible*. DAPP Publicaciones Jurídicas, S.L.

Jacobs, J. (1961). *Muerte y vida de las grandes ciudades*. Madrid: Capitán Swing Libros, S.L. 2011.

Korpivaara, A. (2005). *La experiencia innovadora de Viikki (Helsinki)*. In Ecobarrios en Europa - Nuevos entornos residenciales. Ayuntamiento de Madrid.

Lynch, K. (1984). *La imagen de la ciudad*. Gustavo Gili.

Morán, N. (2008). *Ecobarrio*. Publication on internet. http://habitat.aq.upm.es/temas/a-ecobarrio.html. Accessed 12 July 2016.

Observatorio de Ciudades Inclusivas. (n.d.). *Friburgo, Alemania: Ecobarrio de Vauban*. www.uclg-cisdp.org/sites/default/files/Friburgo_2010_es_FINAL_0.pdf. Accessed 19 July 2016.

Organi-K. (n.d.). Digital journal http://organik.org.mx/nsp/viewpage.php?page_id=3. Accessed 18 July 2016.

Ruano, M. (1999). *Ecourbanismo, entornos humanos sostenibles: 60 proyectos*. Gustavo Gili.

Rudin, D. & Falk, N. (1999). *Building the 21st century home*. The sustainable urban neighbourhood. Great Britain, URBED - Planta Tree.

Rueda Palazuelos, S. (2013). *El urbanismo ecológico*. In urban-e. Territorio, urbanismo, paisaje, sostenibilidad y diseño urbano. Retrieved from http://urban-e.aq.upm el 11/08/2022. Polytechnic Univesity of Madrid.

Verdaguer, C. (2000). *De la sostenibilidad a los ecobarrios*. Universidad Politécnica de Madrid. http://habitat.aq.upm.es/boletin/n14/acver.html#UNO. Accessed 13 July 2016.

References

Aldea Ecológica Huehuecoyotl. (n.d.). http://huehuecoyotl.net. Accessed 18 July 2016.

Arcia, M. E. (2012). *Un barrio de casas ecológicas en Londres*. http://icasasecologicas.com/. Accessed 29 Oct 2012.

Biorregional. (n.d.). *The BedZED Story*. www.bioregional.com/wp-content/uploads/2016/04/The-BedZED-Story.pdf. Accessed 18 July 2016.

Camacho, F. (2010, June 6). Ecobarrios, propuesta ante un panorama donde las megalópolis se colapsan. *La Jornada*. Mexico City. http://www.jornada.unam.mx/2015/06/06/cultura/a03n1cul. Accessed 16 Aug 2016.

Caravana Arcoiris por la Paz. (2012). Ecoaldea Huehuecoyotl, Mexico City. http://caravanaarcoiris.blogspot.mx/. Accessed 16 Aug 2016.

Casa Bioclimática. (2009). Barrios ecológicos. http://www.casabioclimatica.com/noticias/blog/casa-bioclimatica/barrios-ecologicos/. Accessed 20 Aug 2016.

Cátedra Municipios Sostenibles. (2011). *Barrio de Viikki, ciudad de Helsinki, Finlandia*. Universidad Politécnica de Valencia y la Red de Municipios Valencianos hacia la Sostenibilidad http://camuniso.blogspot.mx/2011/01/barrio-de-viikki-ciudad-de-helsinki.html. Accessed 19 July 2016.

Cuello, J. F. (2012). *La verdad sobre los ecobarrios*. http://urbanismosostenible.blogspot.mx/2012/08/la-verdad-sobre-los-ecobarrios.html. Accessed 13 July 2016.

Eroski Consumer. (n.d.). *Barrios ecológicos*. www.consumer.es/web/es/medio_ambiente/urbano/2009/03/18/184097.php?page=3. Accessed 18 July 2016.

Fernández, J. L., Ramos, A., & Morán, N. (2010). De cuartel militar a laboratorio de vida alternativa: el ecobarrio de Vauban en Friburgo. *Electronic journal: Papeles de relaciones ecosociales y cambio global*. E.T.S. Arquitectura, Universidad Politécnica de Madrid. https://raicesyasfalto.wordpress.com/2015/02/01/de-cuartel-militar-a-laboratorio-de-vida-alternativa-el-ecobarrio-de-vauban/. Accessed 19 July 2016.

Hernández, A. A., Velázquez, I., & Verdaguer, C. (2009). *Ecobarrios para ciudades mejores*. Universidad Politécnica de Madrid.

Hieronimi, H. (n.d.). *Huehuecoyotl – una ecoaldea en Morelos/ México*. tierramor.org. http://tierramor.org/Articulos/huehuearticulo.htm. Accessed 9 Jan 2017.

Higueras, E. (n.d.). *El reto de la ciudad habitable y sostenible*. DAPP Publicaciones Jurídicas, S.L.

Jacobs, J. (1961). *Muerte y vida de las grandes ciudades*. Madrid: Capitán Swing Libros, S.L. 2011.

Korpivaara, A. (2005). *La experiencia innovadora de Viikki (Helsinki)*. In Ecobarrios en Europa - Nuevos entornos residenciales. Ayuntamiento de Madrid.

Lynch, K. (1984). *La imagen de la ciudad*. Gustavo Gili.

Morán, N. (2008). *Ecobarrio*. Publication on internet. http://habitat.aq.upm.es/temas/a-ecobarrio.html. Accessed 12 July 2016.

Observatorio de Ciudades Inclusivas. (n.d.). *Friburgo, Alemania: Ecobarrio de Vauban*. www.uclg-cisdp.org/sites/default/files/Friburgo_2010_es_FINAL_0.pdf. Accessed 19 July 2016.

Organi-K. (n.d.). Digital journal http://organik.org.mx/nsp/viewpage.php?page_id=3. Accessed 18 July 2016.

Ruano, M. (1999). *Ecourbanismo, entornos humanos sostenibles: 60 proyectos*. Gustavo Gili.

Rudin, D. & Falk, N. (1999). *Building the 21st century home*. The sustainable urban neighbourhood. Great Britain, URBED - Planta Tree.

Rueda Palazuelos, S. (2013). *El urbanismo ecológico*. In urban-e. Territorio, urbanismo, paisaje, sostenibilidad y diseño urbano. Retrieved from http://urban-e.aq.upm el 11/08/2022. Polytechnic Univesity of Madrid.

Verdaguer, C. (2000). *De la sostenibilidad a los ecobarrios*. Universidad Politécnica de Madrid. http://habitat.aq.upm.es/boletin/n14/acver.html#UNO. Accessed 13 July 2016.